Calculus II

Volumes in the same series:

Calculus I
Complex numbers and differential equations
Matrices and Vectors
Lecturer's Notes

Calculus II

Brian Knight PhD
Senior Lecturer in Mathematics, Goldsmiths' College, London

Roger Adams MSc
Senior Lecturer in Mathematics, Thames Polytechnic, London

London
George Allen & Unwin Ltd
RUSKIN HOUSE · MUSEUM STREET

First published in 1975

This book is copyright under the Berne Convention. All rights are reserved. Apart from any fair dealing for the purpose of private study, research, criticism or review, as permitted under the Copyright Act, 1956, no part of this publication may be reproduced, stored in a retrieval system, or transmitted, in any form or by any means, electronic, electrical, chemical, mechanical, optical, photocopying, recording or otherwise, without the prior permission of the copyright owner. Enquiries should be addressed to the publishers.

© George Allen & Unwin Ltd. 1975.

ISBN 0 04 517012 6

Printed in Great Britain
by William Clowes & Sons, Limited
London, Beccles and Colchester

Introduction

Each chapter in this book deals with a single mathematical topic, which ideally should form the basis of a single lecture. The chapter has been designed as a mixture of the following ingredients:

(i) *Illustrative examples and notes* for the student's pre-lecture reading.
(ii) *Class discussion exercises* for study in a lecture or seminar.
(iii) *Graded problems* for assignment work.

Contents

		page
1	Partial differentiation	11
2	Small increments	16
3	Curvature	21
4	Convergence of series	26
5	Radius of convergence	32
6	Operations with series	37
7	Leibniz' rule	43
8	Integration using trigonometric and hyperbolic functions	47
9	Reduction formulae	55
10	Polar coordinates	59
11	Centre of mass and centroids	65
12	Moments of inertia	70
13	Numerical integration	75
14	Fourier series	81
15	Half-range series	88
	Answers to problems	92
	Index	97

1

Partial Differentiation

In the theory of ordinary differentiation discussed in *Calculus I* we considered the dependent variable y to be a function of a single independent variable x. In many practical applications of the calculus, however, we are often concerned with two, three, or more independent variables. For instance, the power generated in a resistance R in a circuit carrying a current I depends upon both of the variables R and I:

$$P = RI^2$$

It is therefore necessary to extend the rules of differentiation to include functions of several variables of this kind. Here, we shall deal mainly with functions of two variables of the type:

$$f(x, y)$$

and the rules for three or more variables follow a very similar pattern.

Illustrative Example 1

For the function $f(x, y) = x^2y^3 + y^2 - 2x + 7$, if we differentiate f with respect to x treating y as if it were a constant, then we obtain an expression in x and y which is known as the *partial derivative* of f with respect to x and we write:

$$\frac{\partial f}{\partial x} = 2xy^3 - 2$$

For convenience we sometimes write f_x for $\partial f/\partial x$. Similarly:

$$f_y = \frac{\partial f}{\partial y} = 3x^2y^2 + 2y$$

SECOND DERIVATIVES

As the partial derivatives of $f(x, y)$ are functions of x and y, we can differentiate again with respect to x or y to find the second derivatives, written as follows:

$$\frac{\partial^2 f}{\partial x^2} = \frac{\partial}{\partial x}\left(\frac{\partial f}{\partial x}\right) = \frac{\partial}{\partial x}(2xy^3 - 2) = 2y^3 = f_{xx}$$

$$\frac{\partial^2 f}{\partial y \partial x} = \frac{\partial}{\partial y}\left(\frac{\partial f}{\partial x}\right) = \frac{\partial}{\partial y}(2xy^3 - 2) = 6xy^2 = f_{xy}$$

$$\frac{\partial^2 f}{\partial x \partial y} = \frac{\partial}{\partial x}\left(\frac{\partial f}{\partial y}\right) = \frac{\partial}{\partial x}(3x^2y^2 + 2y) = 6xy^2 = f_{yx}$$

$$\frac{\partial^2 f}{\partial y^2} = \frac{\partial}{\partial y}\left(\frac{\partial f}{\partial y}\right) = \frac{\partial}{\partial y}(3x^2y^2 + 2y) = 6x^2y + 2 = f_{yy}$$

We notice that in this example

$$\frac{\partial^2 f}{\partial x \partial y} = \frac{\partial^2 f}{\partial y \partial x}$$

For practically all the cases which we will meet (in fact whenever f_{xy} and f_{yx} are continuous) $f_{xy} = f_{yx}$ and so the order of differentiation is immaterial.

Illustrative Example 2

Find the gradient of the surface $z = x^2y + 1$ at the point (1, 2, 3), (a) when y is kept constant (b) when x is constant.

In the theory of functions of one variable, we pictured the function as a graph of a curve. In terms of two variables, however, we must picture the equation

$$z = x^2y + 1$$

as a surface, shown in figure 1.1. We may see that the equation represents a surface, since to each point Q in the base plane Oxy, there corresponds a single value for z (given by the equation $z = x^2y + 1$). Therefore, as Q moves in the base plane, the point P traces out a surface.

If we now keep y constant, Q may move along the line l, parallel to the x-axis, and P will trace out the curve C. The gradient of this curve is just the derivative of z with respect to x, with y kept constant (see figure 1.1). $\partial z/\partial x$ is the gradient of a 'plane section' of the surface, parallel to the xz-plane. And, similarly: $\partial z/\partial y$ is the gradient of a 'section' of the surface, parallel to the yz-plane.

As an answer to part (a) of the question, we need to find the gradient of a section of the surface through the point (1, 2, 3), and so we simply have to evaluate $\partial z/\partial x$ when $x = 1$, $y = 2$.

$$\frac{\partial z}{\partial x} = 2xy = 4$$

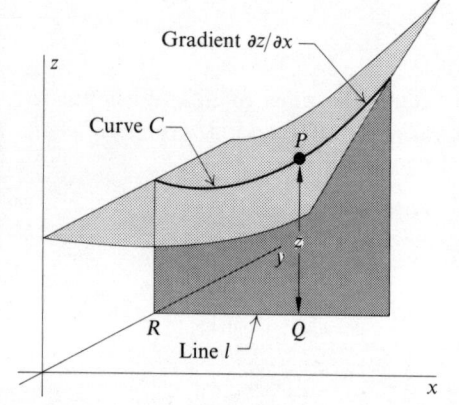

FIGURE 1.1

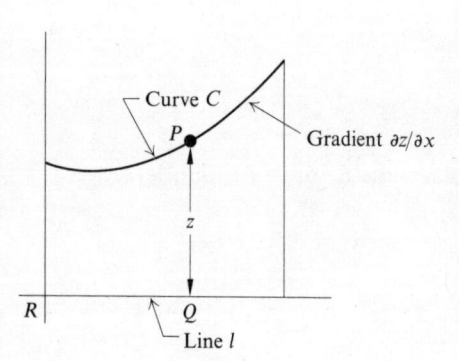

FIGURE 1.2

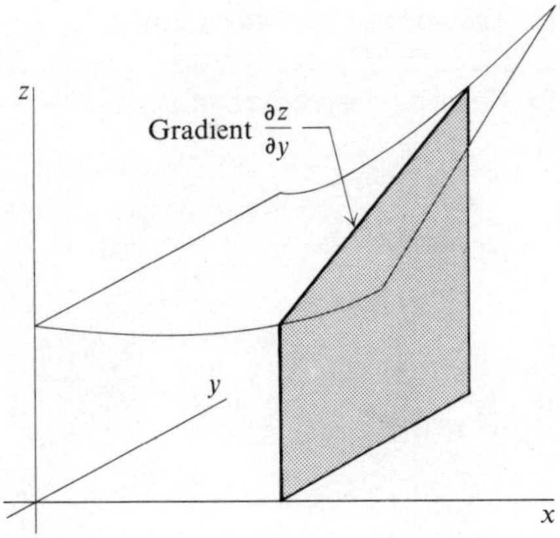

at the required point. To answer part (b), we evaluate $\partial z/\partial y$ at the given point:

$$\frac{\partial z}{\partial y} = x^2 = 1$$

at the point (1, 2, 3).

Class Discussion Exercises 1

1. (i) If $z(x, y) = x^2 + \dfrac{x}{y^2} + 1$, find z_{xx}, z_{yy}, z_{xy}, and z_{yx}

   (ii) If $v = x \log \dfrac{x}{y}$, find $\dfrac{\partial^2 v}{\partial x^2}, \dfrac{\partial^2 v}{\partial y^2}, \dfrac{\partial^2 v}{\partial x \partial y}$

2. If $z(x, y) = \sin \dfrac{x}{y}$, show that:

   (i) $x \dfrac{\partial z}{\partial x} + y \dfrac{\partial z}{\partial y} = 0$ (ii) $x^2 \dfrac{\partial^2 z}{\partial x^2} + 2xy \dfrac{\partial^2 z}{\partial x \partial y} + y^2 \dfrac{\partial^2 z}{\partial y^2} = 0$

3. Find the gradient of the curve in which the surface $z = e^{x(y+1)}$ cuts the plane Oxz, at the point (0, 0, 1).

4. If $z = xf\left(\dfrac{y}{x}\right)$, show that:

   $$x \frac{\partial z}{\partial x} + y \frac{\partial z}{\partial y} = z$$

5B. If $x = r \cos \theta, y = r \sin \theta$, show that x may be written in either forms:

   $$x = r \cos \theta, \quad \text{or} \quad x = \sqrt{(r^2 - y^2)}$$

   Show that the partial derivatives of each of these forms are:

   $$\frac{\partial x}{\partial r} = \cos \theta, \quad \text{and} \quad \frac{\partial x}{\partial r} = \frac{r}{\sqrt{(r^2 - y^2)}} = \frac{1}{\cos \theta}$$

   Explain why these two expressions are different, and justify the notation:

   $$\left(\frac{\partial x}{\partial r}\right)_\theta = \cos \theta, \quad \left(\frac{\partial x}{\partial r}\right)_y = \frac{1}{\cos \theta}$$

Problems 1A

1. Find the indicated partial derivatives in each of the following:

   (i) $V = xy^2 - x^2 + y^2 - 3$; $\dfrac{\partial V}{\partial x}, \dfrac{\partial V}{\partial y}, \dfrac{\partial^2 V}{\partial x^2}, \dfrac{\partial^2 V}{\partial y^2}, \dfrac{\partial^2 V}{\partial x \partial y}, \dfrac{\partial^2 V}{\partial y \partial x}$

FIGURE 1.3

(ii) $V = (x - y)^4$; all first and second derivatives of V
(iii) $z = \sin(xy)$; $z_x, z_y, z_{xx}, z_{xy}, z_{yy}$
(iv) $z = x e^{x-y}$; all first and second derivatives of z
(v) $S = x^2 \cos \dfrac{x}{y}$; $\dfrac{\partial S}{\partial x}, \dfrac{\partial S}{\partial y}$
(vi) $z = uv - \log(uv)$; $z_u, z_v, z_{uu}, z_{uv}, z_{vv}$
(vii) $z = \operatorname{Tan}^{-1}(x/y)$; $\dfrac{\partial z}{\partial x}, \dfrac{\partial z}{\partial y}$
(viii) $f(x, y) = x^2 + y^2$; $f_x(1, 0), f_{xx}(1, 1), f_{xy}(x, y)$
(ix) $z = \dfrac{x}{y} - \dfrac{y}{x}$; all second derivatives of z.
(x) $z = \operatorname{Sin}^{-1} \dfrac{y}{x}$; $\dfrac{\partial z}{\partial x}, \dfrac{\partial z}{\partial y}$

2. If $u = \operatorname{Sin}^{-1} \dfrac{x}{y}$, verify that:
$$x \dfrac{\partial u}{\partial x} + y \dfrac{\partial u}{\partial y} = 0$$

3. If $z = xy \, e^{y/x}$ show that:
   (i) $xz_x + yz_y = 2z$
   (ii) $x^2 z_{xx} + 2xy z_{xy} + y^2 z_{yy} = 2z$

4. Verify that $F(x, y) = \log(x^2 + y^2)$ is a solution of Laplace's equation:
$$\dfrac{\partial^2 F}{\partial x^2} + \dfrac{\partial^2 F}{\partial y^2} = 0$$

5. If $u = x^2 y + y^2 z + z^2 x$, prove that:
   (i) $u_x + u_y + u_z = (x + y + z)^2$
   (ii) $u_{xy} + u_{yz} + u_{zx} = 2(x + y + z)$

6. If $V = \log \dfrac{x}{x - y}$ show that:

(i) $\dfrac{\partial V}{\partial x} + \dfrac{\partial V}{\partial y} = \dfrac{1}{x}$ (ii) $\dfrac{\partial^2 V}{\partial y^2} - \dfrac{\partial^2 V}{\partial x^2} = \dfrac{1}{x^2}$

7. If $u = e^{r \sin \theta}$, show that:
$$r^2 u_{rr} + r u_r + u_{\theta\theta} = r^2 u$$

8. Find the gradient to the curve in which the surface $z = x^2 y^2 - y^2 - x^2 + 1$ cuts the planes:

   (i) Oxz, at the point $(1, 0, 0)$
   (ii) Oyz, at the point $(0, 1, 0)$
   (iii) $y = 2$, at the point $(1, 2, 0)$
   (iv) $x = -1$, at the point $(-1, 1, 0)$

Problems 1B

1. Find $\partial z/\partial x$ and $\partial z/\partial y$ at the point where $x = 1$, $y = 2$ on the surface:
$$x^3 - yz + y^3 = 0$$

2. If $3 + xz + \sin(x - y) = 0$, show that:
$$\frac{\partial z}{\partial x} + \frac{\partial z}{\partial y} = -\frac{z}{x}$$

3. If $V = x^3 f\left(\frac{y}{x}\right)$, show that
$$x^2 V_{xx} + 2xy V_{xy} + y^2 V_{yy} = 6V$$

4. If $z = f(x^2 + y^2)$, prove that:
$$x\frac{\partial z}{\partial y} = y\frac{\partial z}{\partial x}$$

5. If $z = xf\left(\frac{y}{x}\right) + g\left(\frac{x}{y}\right)$, show that:
$$x\frac{\partial z}{\partial x} + y\frac{\partial z}{\partial y} = xf\left(\frac{y}{x}\right)$$

6. Find n if $\theta = t^n \exp\left(\frac{-r^2}{4t}\right)$ satisfies the equation:
$$\frac{\partial}{\partial r}\left\{r^2 \frac{\partial \theta}{\partial r}\right\} = r^2 \frac{\partial \theta}{\partial t}$$

7. If $x = r\cos\theta$, $y = r\sin\theta$, find r and θ in terms of x and y. Find
$\left(\frac{\partial x}{\partial r}\right)_\theta$, $\left(\frac{\partial y}{\partial r}\right)_\theta$, $\left(\frac{\partial x}{\partial \theta}\right)_r$, $\left(\frac{\partial y}{\partial \theta}\right)_r$, and also: $\left(\frac{\partial r}{\partial x}\right)_y$, $\left(\frac{\partial r}{\partial y}\right)_x$, $\left(\frac{\partial \theta}{\partial x}\right)_y$, $\left(\frac{\partial \theta}{\partial y}\right)_x$.

Is
$$\left(\frac{\partial r}{\partial x}\right)_y = 1 \bigg/ \left(\frac{\partial x}{\partial r}\right)_\theta ?$$

Is
$$\left(\frac{\partial r}{\partial y}\right)_x = 1 \bigg/ \left(\frac{\partial y}{\partial r}\right)_\theta ?$$

8. If $x^2 = u + v$ and $y^2 = u - v$, prove that:
$$\left(\frac{\partial x}{\partial u}\right)_v \left(\frac{\partial u}{\partial x}\right)_y = \left(\frac{\partial y}{\partial v}\right)_u \left(\frac{\partial v}{\partial y}\right)_x$$

2
Small Increments

Figure 2.1 illustrates a small section of the surface $z = f(x, y)$. The vertex P has x and y coordinates (x, y), whilst the opposite vertex L has coordinates $(x + \delta x, y + \delta y)$. We may show from this illustration that if x increases by a small amount δx, and y by the small amount δy, then z will increase by:

$$\boxed{\delta z \approx \frac{\partial z}{\partial x} \cdot \delta x + \frac{\partial z}{\partial y} \cdot \delta y}$$

This formula follows from the fact that the gradient of the curve PR is $\partial z/\partial x$, so that $RS \approx \partial z/\partial x \cdot \delta x$, and that the gradient of RL is approximately $\partial z/\partial y$, so that $LM \approx \partial z/\partial y \cdot \delta y$. Hence:

$$\delta z = RS + LM \approx \frac{\partial z}{\partial x} \delta x + \frac{\partial z}{\partial y} \delta y$$

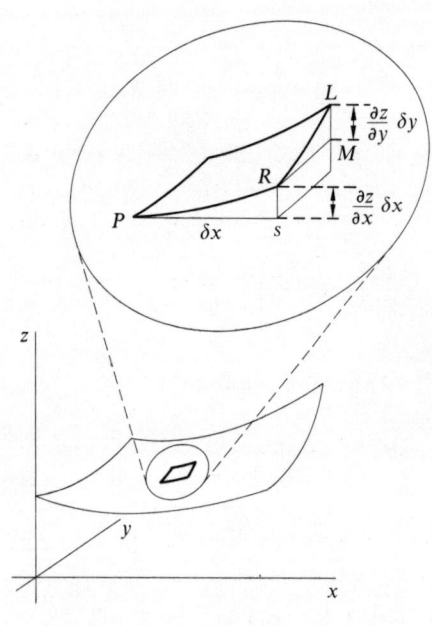

FIGURE 2.1

Illustrative Example 1

The power generated in a resistor R in a circuit carrying a current I is calculated from the formula:

$$P = I^2 R$$

If I is measured as 0·5 ampere, with a maximum error of 0·01 ampere, and R is measured as 50 ohms, with a maximum error of 1 ohm, find the maximum error in the calculated value of P.

Using

$$\delta P \approx \frac{\partial P}{\partial I} \delta I + \frac{\partial P}{\partial R} \delta R$$
$$\delta P \approx 2RI \, \delta I + I^2 \, \delta R$$
$$= 2 \times 0.5 \times 50 \times 0.01 + 0.25 \times 1 = 0.75 \text{ watts}$$

In this case, of course, the formula for δP is relatively simple, and we are also able to calculate δP directly:

$$(P + \delta P) = (I + \delta I)^2(R + \delta R) = I^2R + 2RI\,\delta I + I^2\,\delta R$$

+ terms of second order in small quantities.
Therefore
$$\delta P \approx 2RI\,\delta I + I^2\,\delta R$$

However, in many cases the method by partial differentiation is much quicker.

Class Discussion Exercises 2

1. The volume of a segment of a sphere is given by the formula:

$$V = \frac{1}{6}h(h^2 + 3r^2)$$

where h is the height of a segment and r is the radius of its base. If h is measured as 1 metre with a maximum possible error of 0·005 metre, and r as 2 metres with a maximum error of 0·01 metre, find a good estimate for the maximum error in the computed value of V.

2. The period T in seconds for small oscillations of a compound pendulum is calculated from the formula:

$$T = \frac{2\pi k}{\sqrt{(hg)}}$$

where k is the radius of gyration about the point of oscillation O and h is the distance between O and the centre of gravity. If g is known accurately, but measurements of h and k are subject to errors of ± 1 per cent and ± 2 per cent respectively, show that a good estimate for the maximum error in the calculated value of T is $2\frac{1}{2}$ per cent.

3. If $z = \sin(y/x)$ and $x = t^2, y = 4t$, write z as a function of t, and hence find dz/dt. Show also that dz/dt may be found from the formula:

$$\frac{dz}{dt} = \frac{\partial z}{\partial x}\frac{dx}{dt} + \frac{\partial z}{\partial y}\frac{dy}{dt}$$

Explain how this formula follows from:

$$\delta z = \frac{\partial z}{\partial x}\delta x + \frac{\partial z}{\partial y}\delta y$$

4B. *Total Differential.* The total differential of $z(x, y)$ is defined as dz, where

$$dz = \frac{\partial z}{\partial x}\delta x + \frac{\partial z}{\partial y}\delta y$$

Show that:

(i) $dx = \delta x, dy = \delta y$ (ii) $dz = \dfrac{\partial z}{\partial x} dx + \dfrac{\partial z}{\partial y} dy$

5B. Show that:
$$(2x + y)\, dx + x\, dy$$
is the total differential of a function $f(x, y)$, containing an arbitrary additive constant. Show also that:
$$(2x + 2y)\, dx + x\, dy$$
is not a total differential.

Prove that if $\phi(x, y)\, dx + \psi(x, y)\, dy$ is a total differential, then
$$\frac{\partial \phi}{\partial y} = \frac{\partial \psi}{\partial x}$$

Problems 2A

1. Estimate the maximum error in the calculated value of
$$z = x^2 + 3xy$$
due to errors 0·01 in x and 0·03 in y, which are measured as $x = 2$ and $y = 3$.

2. The phase velocity v of surface waves is given by
$$v = \left(\frac{2\pi T}{\lambda \rho}\right)^{\frac{1}{2}}$$
where T is the surface tension, λ is the wavelength and ρ the density. If λ is known accurately and measurements of T and ρ are correct to within 1 per cent, estimate the maximum percentage error in the calculated value of v.

3. If $f(x, y) = x\, e^{xy}$ and the values of x and y are changed slightly from 1 and 0 to $1 + \delta x$ and δy respectively, so that the change in f is approximately $3\, \delta x$, show $\delta y \approx 2\, \delta x$.

4. Find the approximate change in the hypotenuse of a right-angled triangle of sides 3, 4 and 5 metres, if the 3-metre side is lengthened by 0·25 metre and the 4-metre side is shortened by 0·125 metre.

5. Two sides of a triangle are measured as 15 metres and 20 metres and the included angle as 60°. If the possible errors are 0·2 metre in measuring each of the sides, and 1° in the angle, find an estimate for the maximum error in the computed area.

6. The variance v of the total of claims made out of a portfolio of N policies against an insurance company in one year is computed from the formula:
$$v = N[p\sigma^2 + m^2 p(1 - p)]$$

Here m is the average and σ^2 the variance of the cost of an individual claim. m is estimated at £50 with an error of £0·5, and σ as £10 with an error of £1. The probability that any individual makes a claim is p and is estimated as 0·4 with possible error 0·02. Estimate the maximum possible error in the calculated value of v.

7. If $z = x^2 + 2xy + y^2$, and $x = \sin t$, $y = \cos t$, show that

$$\frac{dz}{dt} = 2(x+y)\cos t - 2(x+y)\sin t$$

8. If $z = \log(x+y)$, $x = e^t$, $y = e^{-t}$, show that $dz/dt = \tanh t$.
9. If $u = xy + yz + zx$, where $x = e^t + e^{-t}$, $y = e^t - e^{-t}$, $z = e^t$, show that

$$\frac{du}{dt} = x(2e^t + e^{-t}) + y(2e^t - e^{-t}) + 2z e^t$$

10. If $z = f(x, y)$, put $t = x$ in the formula of class discussion exercise 2 to show that:

$$\frac{dz}{dx} = \frac{\partial z}{\partial x} + \frac{\partial z}{\partial y}\frac{\partial y}{\partial x}$$

Hence show that:

(i) if $z = x^2 + 2xy + 2y^2$, then

$$\frac{dz}{dx} = 2x + 2y + (2x + 4y)$$

(ii) if $z(x, y) = 0$, then

$$\frac{dy}{dx} = -\frac{\partial z}{\partial x}\bigg/\frac{\partial z}{\partial y}$$

Problems 2B

1. Find the total differential dz in terms of dx and dy for the following:

   (i) $z = x^2 y + y^2 x$ (ii) $z = (x+y)\log\left(\frac{x}{y}\right)$

   (iii) $z = y - y \sin x$ (iv) $z = e^{x+y}$
   (v) $z = y \operatorname{Sinh}^{-1} x$

2. The following expressions are total differentials of a function $f(x, y)$. Find the function f in each case.

   (i) $6x^2 y\, dx + 2x^3\, dy$ (ii) $2(x+y)\, dx + 2(x-y)\, dy$
   (iii) $2x\, dx - 2y\, dy$ (iv) $\sin x\, dy + y \cos x\, dx$

3. Which of the following are total differentials df? Find $f(x, y)$ in the appropriate cases.
   (i) $x^2\, dx + xy\, dy$
   (ii) $(x^2 + \tfrac{1}{2}y^2)\, dx + xy\, dy$
   (iii) $(dx + dy)/(x + y)^2$
   (iv) $(dx - dy)/(x + y)^2$
   (v) $\cos(x + y)(dx + dy)$
   (vi) $(1 - \cos 2x)\, dy + 2y \sin 2x\, dx$

4. The value of V is determined from
$$R^2 + 2EV + E^2 = 0$$

Find the error in V due to small errors δR and δE in R and E. If, additionally, $E + V = R$, show that $2R\,\delta R = V\,\delta V$.

3
Curvature

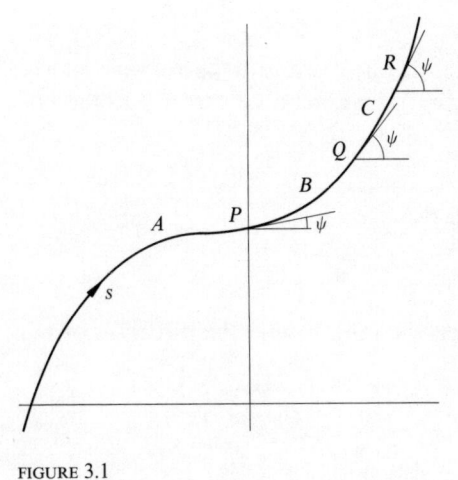

One important measure associated with the graph of a function is that of *curvature*, which roughly speaking measures how curved the graph is. The graph in figure 3.1, for instance, has a high curvature at the point B, and a low curvature at the point C. An idea of the value of the curvature at B is obtained by noticing that the angle ψ, between the tangent to the curve and the direction of the x-axis, increases greatly (by about $\pi/4$) when we pass along the short length of arc PBQ; whereas ψ does not increase by much when we go along an approximately equal length of arc, QCR.

In fact, the idea of the rate of increase of ψ with respect to arc length s is used to define curvature:

$$\kappa = \frac{d\psi}{ds}$$

FIGURE 3.1

Curvature in x, y Coordinates From the definition, $\kappa = d\psi/ds$, we may show (see class discussion exercise 1) that an alternative expression for κ in terms of x and y holds:

$$\kappa = \frac{\dfrac{d^2y}{dx^2}}{+\left\{1 + \left(\dfrac{dy}{dx}\right)^2\right\}^{3/2}}$$

Circle of Curvature The circle touching the curve at a point A, with the same curvature as the curve at A, is called the *circle of curvature* at A. C is called the *centre of curvature*, and ρ the *radius of curvature*. It is shown in class discussion exercise 2 that

$$\rho = \frac{1}{\kappa} = \frac{ds}{d\psi}$$

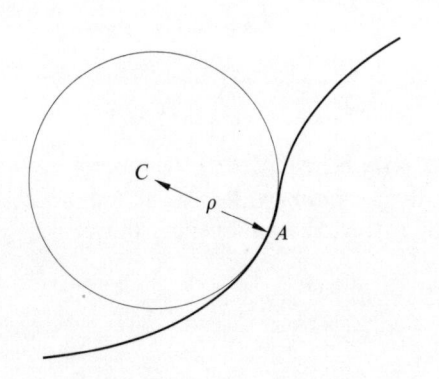

FIGURE 3.2

Curvature

Illustrative Example 1

Find the radius of curvature at the general point on the curve $y = \frac{1}{2}x^2$. Differentiating:

$$y' = x, \quad y'' = 1$$

$$\kappa = \frac{1}{\rho} = \frac{1}{+\{1 + x^2\}^{3/2}}$$

and:

$$\rho = (1 + x^2)^{3/2}$$

Illustrative Example 2

Find the coordinates of the centre of curvature in the example above. From the geometry of figure 3.3, we may see that X, Y, the coordinates of C, are:

$$\boxed{\begin{aligned} X &= x - \rho \sin \psi \\ Y &= y + \rho \cos \psi \end{aligned}}$$

where ψ is the angle between the tangent and the direction of the x-axis.

$$\tan \psi = \frac{dy}{dx} = x$$

Therefore

$$\sec^2 \psi = 1 + \tan^2 \psi = 1 + x^2$$

and

$$\cos \psi = \frac{1}{\sqrt{(1 + x^2)}}, \quad \sin \psi = \frac{x}{\sqrt{(1 + x^2)}}$$

Therefore

$$X = x - (1 + x^2)x = -x^3, \quad \text{and} \quad Y = y + (1 + x^2) = 1 + \tfrac{3}{2}x^2$$

These two equations may if necessary be regarded as the parametric equations of the locus of C (the *evolute*), with x as parameter; whence, on eliminating x, we have the equation of the evolute:

$$Y = 1 + \tfrac{3}{2}X^{2/3}$$

or:

$$8(Y - 1)^3 = 27X^2$$

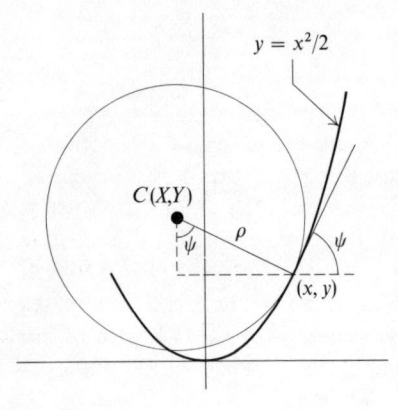

FIGURE 3.3

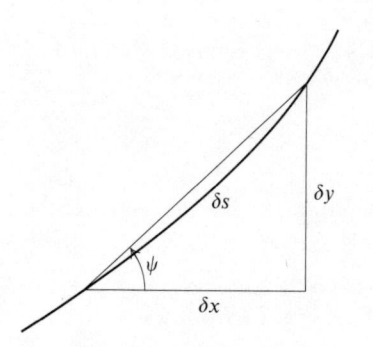

FIGURE 3.4

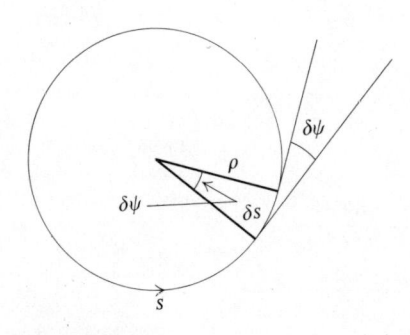

FIGURE 3.5

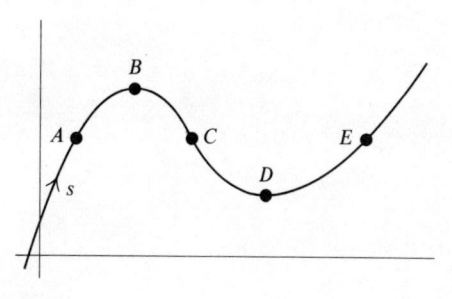

FIGURE 3.6

Class Discussion Exercises 3

1. Find the radius of curvature at the point (1, 1) on the curve:

$$y = x^{3/2}$$

2. Use the diagram of figure 3.4 to show that $dx/ds = \cos\psi$. Hence, by differentiating the equation:

$$\frac{dy}{dx} = \tan\psi$$

with respect to s, show that:

$$\kappa = \frac{d\psi}{ds} = \frac{\dfrac{d^2y}{dx^2}}{+\left\{1 + \left(\dfrac{dy}{dx}\right)^2\right\}^{\frac{3}{2}}}$$

3. Show that for a circle of radius ρ, the curvature is a constant, and equal to $1/\rho$.

Hence show that for a general curve, the radius of curvature is given by:

$$\rho = \frac{1}{\kappa} = \frac{ds}{d\psi}$$

4. Show that with s measured positive from left to right as indicated in figure 3.1, ψ is decreasing as s increases at A, and that the curvature is negative there, whilst it is positive at B.

Show also that the same results may be obtained from the formula of exercise 2. Say whether the curvature is positive, negative, or zero at points A to E in figure 3.6.

5. Show that if a curve is given parametrically by $x = x(t), y = y(t)$, then:

$$\kappa = \frac{1}{\rho} = \frac{\dot{x}\ddot{y} - \ddot{x}\dot{y}}{+(\dot{x}^2 + \dot{y}^2)^{\frac{3}{2}}}$$

Sketch the parabola $(at^2, 2at)$, and show that ρ is negative for positive t. Show that the formulae for the coordinates of the centre of curvature, given in illustrative example 2, apply also for negative ρ. Deduce that

$$\{a(2 + 3t^2), -2at^3\}$$

are the coordinates of the centre of curvature for this example.

Problems 3A

1. Find the radius of curvature for each of the following curves, at the point indicated:

(i) $y = \sqrt{x}$, at $x = 1$
(ii) $y = c \cosh(x/c)$ at $x = 0$
(iii) $y^2 = 1 - \frac{1}{4}x^2$ at the point $(0, 1)$
(iv) $y^2 = x^3$ at $x = 1, y = 1$
(v) $y = 1/x$ at $x = 1$
(vi) $y^3 + x^3 = 2a^3$ at (a, a)

2. Find the radius of curvature of each of the following curves at the general point, (x, y):

(i) $y = \log(\cos x)$ (ii) $y = \mathrm{Sin}^{-1} x$
(iii) $y = \log(\sin x)$ (iv) $y = \frac{1}{4}x^2 - \frac{1}{2}\log x$

3. Find the radius of curvature of each of the following curves at the general point:

(i) $x = ct, y = c/t$
(ii) $x = 2\cos\theta, y = \sin\theta$
(iii) $x = a\cos^3 t, y = a\sin^3 t$
(iv) $x = a\cos t^2, y = a\sin t^2$
(v) $x = 2a\sin t + a\sin 2t, y = 2a\cos t + a\cos 2t$
(vi) $x = a(\cos\theta + \theta\sin\theta), y = a(\sin\theta - \theta\cos\theta)$
(vii) $x = a(t + \sin t), y = a(1 + \cos t)$

4. Show that a straight line has zero curvature.

5. Show that the centre of curvature at the point $(1, 1)$ on the curve

$$y = \frac{1}{x}$$

has coordinates $(2, 2)$.

Problems 3B

1. Show that the radius of curvature for the curve:

$$x = a(\cos t + t\sin t), \qquad y = a(\sin t - t\cos t)$$

is at. If $\tan\psi = dy/dx$, show that $\psi = t$. Hence show that the coordinates of the centre of curvature are:

$$X = a\cos t, \qquad Y = a\sin t$$

2. Show that the radius of curvature for the curve:

$$x = a\cos^3 t, \qquad y = a\sin^3 t$$

is $3a\sin t \cos t$. If $\tan\psi = dy/dx$, show that $\psi = t - \frac{1}{2}\pi$. Show that the equation of the evolute is:

$$X = a\cos^3 t + 3a\sin^2 t \cos t$$
$$Y = a\sin^3 t + 3a\sin t \cos^2 t$$

3. Show that the radius of curvature at the point t on the curve
$$x = a(t - \sin t), \qquad y = a(1 - \cos t)$$
is $-4 \sin(\tfrac{1}{2}t)$. Hence show that the evolute is given by:
$$X = a(t + \sin t), \quad Y = -a(1 - \cos t)$$

4
Convergence of Series

Infinite series have already been discussed to some extent in *Calculus I*, although as yet we have not clearly defined what is meant by a sum of the form:

$$s = \sum_{n=1}^{\infty} u_n = u_1 + u_2 + \cdots + u_n + \cdots$$

In order to do this, we first define a partial sum of the series as:

$$s_n = u_1 + u_2 + \cdots + u_n$$

Now the sum s may be defined as:

$$s = \lim_{n \to \infty} s_n$$

when the limit exists. If the limit does exist the series is said to be *convergent* or *summable*, if it does not exist then the series is *divergent* or *not summable*.

Illustrative Example 1

For the infinite geometric progression:

$$\sum_{n=1}^{\infty} r^n = 1 + r + r^2 + \cdots$$

we have a partial sum:

$$s_n = 1 + r + r^2 + \cdots + r^n = \frac{1 - r^{n-1}}{1 - r}$$

If $r < 1$, $\lim_{n \to \infty} s_n = 1/(1 - r)$, and the series is convergent with sum $1/(1 - r)$. If $r > 1$, $\lim_{n \to \infty} s_n = \infty$, and since the limit does not exist the series is divergent.

Illustrative Example 2

Show that the following series is convergent:

$$2^{100} + 2^{99} + 2^{98} + \cdots + 1 + \tfrac{1}{2} + \tfrac{1}{4} + \tfrac{1}{8} + \cdots$$

In this series,

$$u_1 = 2^{100}, u_2 = 2^{99}, \ldots, u_{100} = 1, \ldots$$

and we have:

$$\sum_{n=100}^{\infty} u_n = 1 + \tfrac{1}{2} + \tfrac{1}{4} + \tfrac{1}{8} + \cdots$$

This series converges to the sum $1/(1 - \tfrac{1}{2}) = 2$. Hence the whole series must have the large but finite sum:

$$\sum_{n=1}^{\infty} u_n = 2^{100} + 2^{99} + \cdots + 2 + (2)$$

and is also convergent. We see from this example that the convergence of a series depends only upon its behaviour for large n. If the series $\sum_{n=N}^{\infty} u_n$ is convergent, then so is $\sum_{n=1}^{\infty} u_n$, where N is any large finite number.

Another helpful rule follows by noticing that if $\sum_{n=1}^{\infty} u_n$ converges to the sum s, then $\sum_{n=1}^{\infty} cu_n$ will converge to the sum cs. Thus we may multiply the terms of a convergent series by any finite constant and obtain another convergent series.

THE COMPARISON TEST FOR SERIES OF POSITIVE TERMS

> If $0 \leq u_n \leq v_n$ for all n, then
> if $\sum_{n=1}^{\infty} v_n$ converges, so does $\sum_{n=1}^{\infty} u_n$

Illustrative Example 3

Is the series

$$\sum_{n=1}^{\infty} \frac{1}{2^n - n + 1}$$

convergent? In practice we may see at a glance that this series converges, since as we know from illustrative example 2 convergence depends only upon the behaviour for large n, and the nth term of this series is practically $1/2^n$ for large n. To be precise:

$$\lim_{n \to \infty} \frac{\left(\dfrac{1}{2^n - n + 1}\right)}{\left(\dfrac{1}{2^n}\right)} = 1$$

Therefore if we choose $c > 1$, it is certainly true that:
$$0 \leqslant \frac{1}{2^n - n + 1} \leqslant c \frac{1}{2^n}$$
for large enough n (i.e. for all $n > N$). Since
$$\sum_{n=N}^{\infty} c \frac{1}{2^n}$$
is a convergent geometric progression, therefore by the comparison test
$$\sum_{n=N}^{\infty} \frac{1}{2^n - n + 1}$$
is also convergent. The convergence of
$$\sum_{n=1}^{\infty} \frac{1}{2^n - n + 1}$$
follows by the result of illustrative example 2.

THE RATIO TEST FOR A SERIES OF POSITIVE TERMS

> If $u_n > 0$ for all n and
> $$\lim_{n \to \infty} \frac{u_{n+1}}{u_n} = r$$
> then
> $$\sum_{n=1}^{\infty} u_n \quad \begin{array}{l} \text{converges if } r < 1 \\ \text{diverges if } r > 1 \end{array}$$

Illustrative Example 4

Investigate the convergence of the series $\sum_{n=1}^{\infty} 1/n!$ Here we have
$$u_n = \frac{1}{n!}, \quad u_{n+1} = \frac{1}{(n+1)!}$$
Therefore
$$\frac{u_{n+1}}{u_n} = \frac{1/(n+1)!}{1/n!} = \frac{n!}{(n+1)!} = \frac{1}{n+1}$$
Thus
$$\lim_{n \to \infty} \frac{u_{n+1}}{u_n} = 0$$
In this case $r = 0$ and is therefore less than 1. Hence the series converges (to the sum e).

Class Discussion Exercises 4

1. Use the comparison test to show that
$$\sum_{n=1}^{\infty} \frac{2}{3^n - n^2 + n}$$
is convergent.

2. Use the ratio test to show that the following series converges for all x:
$$1 + \frac{x^2}{2!} + \frac{x^4}{4!} + \cdots$$

3. Show that the harmonic series:
$$\sum_{n=1}^{\infty} \frac{1}{n} = 1 + \tfrac{1}{2} + \tfrac{1}{3} + \tfrac{1}{4} + \cdots$$
is divergent. In fact this is a critical case of a more general result proved in 7B, which is of great use in conjunction with the comparison test:

> The series $\sum_{n=1}^{\infty} 1/n^p$ is convergent if $p > 1$ and divergent if $p \leq 1$

Use this result to investigate the convergence of:

(i) $\sum_{n=1}^{\infty} \frac{n+1}{n^2 + 5}$ (ii) $\sum_{n=1}^{\infty} \frac{n-1}{n^3 + 5}$ (iii) $\sum_{n=2}^{\infty} \frac{\sqrt{(n+1)}}{2n - 1}$

4. Show that if $\sum_{n=1}^{\infty} u_n$ is convergent then $\lim_{n \to \infty} u_n = 0$.

5. *Leibniz' theorem.* Show that if $a_1 \geq a_2 \geq a_3 \ldots \geq 0$ and $\lim_{n \to \infty} a_n = 0$, then the series:
$$\sum_{n=1}^{\infty} u_n = \sum_{n=1}^{\infty} (-1)^{n+1} a_n = a_1 - a_2 + a_3 - a_4 + \cdots$$
converges. Hence show that the following are convergent:

(i) $1 - \frac{1}{2!} + \frac{1}{4!} - \cdots$ (ii) $1 - \tfrac{1}{2} + \tfrac{1}{3} - \tfrac{1}{4} + \cdots$

6. *Absolute Convergence.* If the series $\sum_{n=1}^{\infty} |u_n|$ is convergent, $\sum_{n=1}^{\infty} u_n$ is said to be *absolutely convergent*. It may be shown that if a series is absolutely convergent, then it is convergent. State which of the series in exercise 5 is absolutely convergent.

7B. Prove (i) the comparison test (ii) the ratio test and (iii) the result stated in exercise 3 for the series $\sum_{n=1}^{\infty} 1/n^p$.

Problems 4A

Investigate the convergence of the following series:

1. (i) $\sum_{n=1}^{\infty} \dfrac{2 + 100n}{n^3 + n - 1}$ (ii) $\sum_{n=1}^{\infty} \dfrac{2^n + n}{3^n - 1}$ (iii) $\sum_{n=100}^{\infty} \dfrac{2}{\sqrt{n}}$

   (iv) $\sum_{n=1}^{\infty} \dfrac{n^3 + n}{100n^2 + 1}$ (v) $\sum_{n=1}^{\infty} \dfrac{n^2 + \cos n}{3n^4 - n}$ (vi) $\sum_{n=1}^{\infty} \dfrac{1}{\sqrt[3]{(n^2 + 1)}}$

2. (i) $\sum_{n=1}^{\infty} \dfrac{1}{(2n + 1)!}$ (ii) $\sum_{n=1}^{\infty} \dfrac{n^2}{n!}$

   (iii) $\sum_{n=1}^{\infty} \dfrac{2^n}{n^3}$ (iv) $\sum_{n=1}^{\infty} \dfrac{10^n}{n!}$

3. (i) $\tfrac{1}{2} - 2(\tfrac{1}{2})^2 + 3(\tfrac{1}{2})^3 - 4(\tfrac{1}{2})^4 + \cdots$

   (ii) $2 - \dfrac{2^2}{2} + \dfrac{2^3}{3} - \dfrac{2^4}{4} + \cdots$

   (iii) $\sum_{n=1}^{\infty} (-1)^n \dfrac{\log n}{n}$

   (iv) $\sum_{n=1}^{\infty} (-1)^n \dfrac{1}{\sqrt{n}}$

4. Which of the series in problem 3 are absolutely convergent?

5. For what values of x are the following series convergent?

   (i) $-x + \tfrac{1}{2}x^2 - \tfrac{1}{3}x^3 + \tfrac{1}{4}x^4 - \cdots$

   (ii) $1 + x + \dfrac{x^2}{2!} + \dfrac{x^3}{3!} + \cdots$

   (iii) $x - \dfrac{x^2}{2!} + \dfrac{x^4}{4!} - \cdots$

Problems 4B

1. Show that the series

$$\sum_{n=1}^{\infty} \dfrac{a^{2n}}{(1 + a^2)^{n-1}}$$

   is convergent for all finite a.

2. Use the ratio test to investigate the convergence of:

   (i) $\sum_{n=1}^{\infty} \dfrac{n!}{n^n}$ (ii) $\sum_{n=1}^{\infty} \dfrac{2^n n!}{n^n}$ (iii) $\sum_{n=1}^{\infty} \dfrac{3^n n!}{n^n}$

3. (i) Show that $\sum_{n=2}^{\infty} v_n$ is convergent, where

$$v_n = \int_{n-1}^{n} \frac{1}{x(\log x)^2} dx$$

Hence show by the comparison test that

$$\sum_{n=1}^{\infty} \frac{1}{n(\log n)^2}$$

is convergent.

(ii) Show that $\sum_{n=2}^{\infty} v_n$ is divergent, where

$$v_n = \int_{n-1}^{n} \frac{1}{x \log x} dx$$

Hence show that

$$\sum_{n=1}^{\infty} \frac{1}{n \log n}$$

is divergent.

(iii) Test the series $\sum_{n=1}^{\infty} \log n/n$ for convergence.

4. Use the inequality: $(|a_n| - |b_n|)^2 \geq 0$ to show that if $\sum a_n^2$ and $\sum b_n^2$ are convergent, then $\sum a_n b_n$ is absolutely convergent.

5
Radius of Convergence

Using the theory of convergence of series discussed in Chapter 4, we may now show that the general power series expansion:

$$f(x) = a_0 + a_1 x + a_2 x^2 + a_3 x^3 + \cdots$$

is valid for $|x| < R$, and not valid for $|x| > R$, where R is known as the *radius of convergence* of the series. In order to find R for a particular series, we use the ratio test to find which values of x make the series absolutely convergent.

Illustrative Example 1

Find the radius of convergence of the power series:

$$\log_e (1 - 2x) = -2x - \frac{2^2 x^2}{2} - \frac{2^3 x^3}{3} - \cdots - \frac{2^n x^n}{n} - \cdots$$

$$\frac{|u_{n+1}|}{|u_n|} = \left| \frac{2^{n+1} x^{n+1}}{n+1} \middle/ \frac{2^n x^n}{n} \right| = \left| \frac{2xn}{n+1} \right|$$

Therefore

$$\lim_{n \to \infty} \frac{|u_{n+1}|}{|u_n|} = 2|x|$$

Hence the series is absolutely convergent and therefore convergent for $|x| < \frac{1}{2}$, i.e. $R = \frac{1}{2}$. The ratio test also shows that the series is divergent if $|x| > \frac{1}{2}$, but gives no information as to the convergence for the cases $x = \pm R$. We consider these 'end-points' in class discussion exercise 5.

Illustrative Example 2

It is often useful when we are making approximate calculations to take only the first few terms in the Maclaurin series of a function. In this case we need to be able to estimate the maximum possible error which we are making. If the power series under consideration is:

$$a_0 + a_1 x + \cdots + a_{n-1} x^{n-1} + a_n x^n + \cdots$$

then the error involved in using only the first n terms of the series (that is up to the term in x^{n-1}) is:

$$R_n = a_n x^n + a_{n+1} x^{n+1} + \cdots$$

For example if we take $x = 0.1$ in the series for $\log_e (1 - 2x)$ we obtain:

$$\log_e (0.8) = 0 - 0.2 - 0.02 - 0.002667 - 0.0004 - 0.000069 - \cdots$$

Hence if we use the first three terms in the series we obtain:

$$\log_e (0.8) = -0.22$$

and the error is:

$$R_3 = -0.002667 - 0.0004 - 0.000069 - \cdots \approx -0.003144$$

Notice that the terms in this series are becoming small quite quickly. This is because $x = 0.1$ is quite small compared with the radius of convergence $R = 0.5$. Consequently the error: $R_3 = a_3 x^3 + a_4 x^4 + \ldots$ is dominated by the first term $a_3 x^3 = 0.002667$. Hence a useful guide is that:

> The error in curtailing a series is the same order of magnitude as the next term in the series, whenever x is small compared with R.

We may see how well this rule works by examining the following table relating to the series for $\log_e (1 - 2x)$, when $x = 0.1$:

n	2	3	4	5
R_n (error after term in x^{n-1})	0.02314	0.00314	0.00048	0.00008
$\|a_n x^n\|$ (next term)	0.02000	0.00267	0.00040	0.00006

The smaller x becomes the more accurate the rule becomes, as we see from the following table when $x = 0.01$:

n	2	3	4	5
R_n	2.0×10^{-4}	2.7×10^{-6}	4.1×10^{-8}	6.5×10^{-10}
$\|a_n x^n\|$	2.0×10^{-4}	2.7×10^{-6}	4.0×10^{-8}	6.4×10^{-10}

However, if x is close to R then the rule does not give such a good estimate of the error, e.g. when $x = 0.4$ we have:

n	2	3	4	5
R_n	0·809	0·489	0·319	0·216
$|a_n x^n|$	0·320	0·170	0·102	0·066

Class Discussion Exercises 5

1. Find the radius of convergence for the binomial series $(1 + x)^k$ when k is not equal to a positive integer.
2. Write down the general term in the following series and find the radius of convergence in each case:

(i) $\dfrac{x}{2} + \dfrac{2x^2}{2^2} + \dfrac{3x^3}{2^3} + \dfrac{4x^4}{2^4} + \cdots$

(ii) $\dfrac{x}{1 \times 2} - \dfrac{x^2}{2 \times 3} + \dfrac{x^3}{3 \times 4} - \dfrac{x^4}{4 \times 5} + \cdots$

(iii) $2x^2 + \dfrac{2^2 x^4}{2} + \dfrac{2^3 x^6}{3} + \cdots$

(iv) $x - \dfrac{x^3}{3!} + \dfrac{x^5}{5!} - \cdots$

3. Compute $e^{0.2}$ from the first three terms in the series for e^x and estimate the error. Given that $e^{0.2} = 1.2214$ calculate the actual error and compare it with your estimate. What is the radius of convergence of the series?
4. How many terms in the series for $\log_e (1 + x)$ must we take in order to calculate $\log_e 1.5$ with an error less than 10^{-3}? Write down the series expansion for $\log_e (1 + x)/(1 - x)$. How many terms are needed in this series to calculate $\log_e 1.5$ with an error less than 10^{-3}?
5. Is the series expansion of $\log_e (1 - 2x)$ valid for: (i) $x = \tfrac{1}{2}$, (ii) $x = -\tfrac{1}{2}$?

Problems 5A

1. Find the radius of convergence of the following series:

(i) $\sum\limits_{n=1}^{\infty} (-1)^n n x^{n-1}$ (ii) $\sum\limits_{n=1}^{\infty} \dfrac{(-2)^n x^n}{n(n+1)}$

2. Write down the general term in the following series and find the radius of convergence in each case:

(i) $1 + \dfrac{1.2}{1.3}x + \dfrac{1.2.3}{1.3.5}x^2 + \dfrac{1.2.3.4}{1.3.5.7}x^3 + \cdots$

(ii) $1 - \dfrac{x}{2.2} + \dfrac{x^2}{2^2.3} - \dfrac{x^3}{2^3.4} + \cdots$

3. Use the series for $(1 + x)^{-1}$ to the number of terms indicated in order to compute the following and estimate the size of the remainders in each case:

(i) $(1{\cdot}02)^{-1}$ (2 terms) (ii) $(1{\cdot}02)^{-1}$ (3 terms)
(iii) $(0{\cdot}98)^{-1}$ (3 terms) (iv) $(0{\cdot}99)^{-1}$ (3 terms)

4. Compute $\log_e (0{\cdot}95)$ from the series for $\log_e (1 - x)$ using the terms up to and including the term in x^3. Estimate the remainder.

5. Find the radius of convergence of the following series:

(i) $\sum\limits_{n=0}^{\infty} \dfrac{x^{2n+1}}{(2n + 1)}$ (ii) $\sum\limits_{n=0}^{\infty} \dfrac{n^3 x^{2n}}{9^n}$ (iii) $\sum\limits_{n=0}^{\infty} \dfrac{x^{4n}}{3^n}$

6. Write down the general term in the following series and find the radius of convergence in each case:

(i) $1 + \dfrac{x^2}{2.1} + \dfrac{x^4}{2^2.2} + \dfrac{x^6}{2^3.3} + \cdots$

(ii) $x + \dfrac{1.2}{1.3}x^3 + \dfrac{1.2.3}{1.3.5}x^5 + \dfrac{1.2.3.4}{1.3.5.7}x^7 + \cdots$

7. Write down the Maclaurin series for $\sin x$.
Calculate $\sin 1$ using: (i) the first two, (ii) the first three, (iii) the first four non-zero terms. Estimate the error in each case.

8. How many terms are necessary if we require an absolute error not exceeding 10^{-6} in computing $\log_e (1 + x)$, where $0 \leq x \leq 0{\cdot}1$?

9. Estimate the error involved in making the approximation:

$\sqrt{(1 + x)} = 1 + \tfrac{1}{2}x$ for values of x such that $|x| < 10^{-3}$

10. Calculate $e^{0 \cdot 1}$ correct to five decimal places.

11. Show that
$$\dfrac{3x}{9 + x^2} = \sum_{n=0}^{\infty} (-1)^n \left(\dfrac{x}{3}\right)^{2n+1}$$
and find the radius of convergence of the series.

Problems 5B

1. By putting $x + 1 = x'$, show that
$$\frac{1}{x} = -\sum_{n=0}^{\infty} (1 + x)^n$$
and find values of x for which the series is convergent.

2. Find a range of values of x, in the form $a < x < b$, for which each of the following series converges:

   (i) $\sum_{n=0}^{\infty} \frac{(x + 1)^n}{(n + 1)2^n}$

   (ii) $\sum_{n=0}^{\infty} \frac{(x + 1)^n}{(n + 1)^2}$

   (iii) $\sum_{n=0}^{\infty} \frac{(x - 2)^{2n}}{(n + 1)(-1)^n}$

   (iv) $\sum_{n=1}^{\infty} n^n (x - 1)^n$

   (v) $\sum_{n=1}^{\infty} \frac{n}{(n + 1)} \left(\frac{2 + x}{2}\right)^n$

3. Use the first two terms of a series to compute $\sin 46°$ and estimate the size of the error.

4. Calculate the maximum error in taking:
$$\sin x = x - \frac{x^3}{6} + \frac{x^5}{120} \quad \text{when } |x| \leq \frac{\pi}{2}$$

5. Taking $(1 + x)^{-1} = 1 - x$, what range of positive values of x will give an error of less than $0\cdot 01$? Give an estimate of the error when $x = -0\cdot 01$. Compute the error directly.

6. Prove that
$$\log_e (10 + x) = \log_e 10 + x/10 - x^2/200 + x^3/3000 - \cdots$$
if $-10 < x < 10$. Hence find $\log_e 12$ to four decimal places given that $\log_e 10 = 2\cdot 30259$.

7. The relation between the pressure p, and the density ρ, of air during a sound vibration is given by $\rho = kp^{\frac{3}{5}}$, where k is a constant. Denoting the density corresponding to a pressure of one atmosphere by ρ_0 show, by considering the first two terms of the Taylor expansion of ρ as a series of powers of $(p - 1)$, that
$$\rho \approx \frac{\rho_0}{5}(2 + 3p)$$
Show that if p deviates by no more than 1 per cent from 1 atmosphere then the error in taking the above approximate expression for ρ is less than $1\cdot 2 \times 10^{-5}$.

6
Operations with Series

We give here a table of the more important elementary series for reference:

$$e^x = \sum_{n=0}^{\infty} \frac{x^n}{n!} \qquad \text{All } x$$

$$\sin x = \sum_{n=0}^{\infty} \frac{(-1)^n x^{2n+1}}{(2n+1)!} \qquad \text{All } x$$

$$\cos x = \sum_{n=0}^{\infty} \frac{(-1)^n x^{2n}}{(2n)!} \qquad \text{All } x$$

$$\tan x = x + \frac{x^3}{3} + \frac{2x^5}{15} + \frac{17x^7}{315} + \cdots \qquad |x| < \frac{\pi}{2}$$

$$\log_e (1 + x) = \sum_{n=1}^{\infty} \frac{(-1)^{n-1} x^n}{n} \qquad -1 < x \leqslant 1$$

$$(1 + x)^k = \sum_{n=0}^{\infty} \frac{k(k-1)\ldots(k-n+1)x^n}{n!} \qquad |x| < 1$$

(This series is the *binomial series*, valid for all values of x if k is a positive integer, when the series reduces to a polynomial.)

$$\operatorname{Tan}^{-1} x = \sum_{n=1}^{\infty} \frac{(-1)^{n-1} x^{2n-1}}{2n-1} \qquad |x| \leqslant 1$$

$$\sinh x = \sum_{n=0}^{\infty} \frac{x^{2n+1}}{(2n+1)!} \qquad \text{All } x$$

$$\cosh x = \sum_{n=0}^{\infty} \frac{x^{2n}}{(2n)!} \qquad \text{All } x$$

$$\tanh x = x - \frac{x^3}{3} + \frac{2x^5}{15} - \frac{17x^7}{315} + \cdots \qquad |x| < \frac{\pi}{2}$$

$$\tanh^{-1} x = \sum_{n=1}^{\infty} \frac{x^{2n-1}}{2n-1} \qquad |x| < 1$$

Illustrative Example 1: Differentiation and Integration of Series

The radius of convergence of the series:

$$\log_e (1 + x) = x - \frac{x^2}{2} + \frac{x^3}{3} - \cdots + (-1)^{n+1} \frac{x^n}{n} + \cdots$$

is $R = 1$. Differentiating throughout with respect to x, we obtain:

$$1/(1 + x) = 1 - x + x^2 - x^3 + \cdots + (-1)^n x^n + \cdots$$

i.e. a binomial series. Notice that the radius of convergence of the differentiated series is also 1, thus illustrating the following general rule:

> In general a series of powers of x with radius of convergence R may be differentiated or integrated term by term, and the radius of convergence of the resulting series is also R.

Illustrative Example 2: Multiplication of Series

> Two series with radii of convergence R_1 and R_2 may be multiplied together and the resulting series will have radius of convergence R, equal to the smaller of R_1 and R_2.

Write down the first three terms in the series expansion for $f(x) = \sqrt{(1 - 2x)} \sin x$.

In this case, direct application of Maclaurin's series proves to be difficult because of the complexity of the differentiation. However, the series for $\sin x$ and $\sqrt{(1 - 2x)}$ are elementary, and we may write:

$$f(x) = \left(1 - x - \frac{x^2}{2} + \cdots\right)\left(x - \frac{x^3}{3!} + \frac{x^5}{5!} + \cdots\right)$$

$$= \left(x - \frac{x^3}{3!} + \cdots\right) - x\left(x - \frac{x^3}{3!} + \cdots\right)$$

$$\quad - \frac{x^2}{2}\left(x - \frac{x^3}{3!} \cdots\right) + \text{terms in } x^4 \text{ and higher}$$

Hence, calculating only those terms up to powers of x^3:
$$\sqrt{(1-2x)}\sin x = x - x^2 - \tfrac{2}{3}x^3 + \cdots$$

Class Discussion Exercises 6

1. By considering the series for $(1-x^2)^{-\frac{1}{2}}$, find a series to represent $\mathrm{Sin}^{-1} x$. What is the radius of convergence of this series?
2. Use the relation $\int \sin x \, dx = -\cos x + c$ to deduce the $\cos x$ series from the series for $\sin x$.
3. Find the first three terms in the series for $e^x \sin x$.
4. Find a series up to the term in x^4 for (i) $e^{\sin x}$, (ii) $e^{\cos x}$.
5. By dividing the series for $\sin x$ by the series for $\cos x$, find a series for $\tan x$ up to the term in x^5.
6. Evaluate:
$$\lim_{x \to 0} \left\{ \frac{x^2 e^{\sin x} - x^2}{\mathrm{Sin}^{-1} x - x} \right\}$$

7B. Evaluate $\int_0^{0.2} e^{\sin x} dx$, to four decimal places.

Problems 6A

1. Derive the series for $\sinh x$ given that
$$\cosh x = 1 + \frac{x^2}{2!} + \frac{x^4}{4!} + \frac{x^6}{6!} + \cdots$$
   (i) by differentiating this series term by term,
   (ii) by integrating this series term by term.

2. By using the binomial theorem, find the first four terms in the series expansions of
$$\frac{1}{1+x}, \quad \frac{1}{1-x}, \quad \frac{1}{1-x^2}$$
stating for which values of x each expansion is valid. By integrating each term by term, obtain series for $\log_e (1+x)$, $\log_e (1-x)$, $\mathrm{Tanh}^{-1} x$. Hence obtain the identity
$$\mathrm{Tanh}^{-1} x = \tfrac{1}{2} \log_e \left(\frac{1+x}{1-x} \right)$$

3. Show that
$$(1+t^2)^{-1} = 1 - t^2 + t^4 - t^6 + \cdots + (-1)^n t^{2n} + \cdots$$
By integrating from 0 to x, obtain an expansion for $\mathrm{Tan}^{-1} x$.

Hence show that

$$\frac{\pi}{4} = 1 - \tfrac{1}{3} + \tfrac{1}{5} - \tfrac{1}{7} + \cdots$$

4. Find the series expansion of $\log_e (1 + \sin x)$ as far as the term in x^4. For what values of x is this expansion valid?

5. Show that
   (i) $e^{-x} \cos x = 1 - x + \dfrac{x^3}{3} - \cdots$
   (ii) $\cos x \cosh x = 1 - \dfrac{x^4}{6} + \cdots$

6. Obtain the Maclaurin expansion of $(1 + \sin x)^n$ as far as the term in x^3.

7. Show that:
   (i) $\cos^2 x = 1 - x^2 + \tfrac{1}{3}x^4 - \tfrac{2}{45}x^6 + \cdots$
   (ii) $\log_e (1 + e^x) = \log_e 2 + \tfrac{1}{2}x + \tfrac{1}{8}x^2 - \tfrac{1}{192}x^4 + \cdots$
   (iii) $\log_e (\cosh x) = \tfrac{1}{2}x^2 - \tfrac{1}{12}x^4 + \tfrac{1}{45}x^6 - \cdots$

8. Find the value of the following limits by expressing the functions involved as a series of ascending powers

   (i) $\lim\limits_{x \to 0} \dfrac{1 - e^{-x}}{x}$
   (ii) $\lim\limits_{x \to 0} \dfrac{\sin x - x \cos x}{x^3}$

   (iii) $\lim\limits_{x \to 0} \dfrac{x - \sin x}{\tan x - x}$
   (iv) $\lim\limits_{x \to 0} \dfrac{\sin x - x}{\sin x - x \cos x}$

   (v) $\lim\limits_{m \to 0} \left\{ \dfrac{A}{m^2} \left[\dfrac{\sin mt}{\sin ml} - \dfrac{t}{l} \right] \right\}$
   (vi) $\lim\limits_{x \to 0} \dfrac{\text{Tanh}^{-1} x - \tanh x}{x^3}$

   (vii) $\lim\limits_{\omega \to 2} \dfrac{\cos \omega t - \cos 2t}{4 - \omega^2}$
   (viii) $\lim\limits_{x \to 0} \dfrac{x e^{x^2} - \tanh x}{x^3}$

   (ix) $\lim\limits_{x \to 0} \dfrac{(1 - x^2)^{-\frac{1}{2}} - (1 - x^2)^{\frac{1}{2}}}{x^2}$

9. Show that

$$\log_e (\cos x) = - \left[\dfrac{x^2}{2} + \dfrac{x^4}{12} + \dfrac{x^6}{45} + \cdots \right]$$

and hence evaluate $\log_e \{\sec (0 \cdot 2)\}$ correct to four decimal places.

10. By dividing the series for $\sinh x$ by the series for $\cosh x$, find a series for $\tanh x$ up to the term in x^5.

Problems 6B

1. Find the series expansion of $x/\sinh x$ as far as the term in x^4.

Hence evaluate:
$$\lim_{x \to 0} \frac{1}{x^4}\left(\frac{x}{\sinh x} + \frac{\sinh x}{x} - 2\right)$$

2. Prove that
$$\log\{\sec(\pi/4 + x)\} = \tfrac{1}{2}\log 2 + x + x^2 + \tfrac{2}{3}x^3 + \tfrac{2}{3}x^4 + \cdots$$

3. The current i in a circuit containing an inductance L, a capacitance C and an alternator of angular frequency ω is given by
$$i = \frac{\omega E_m}{L(n^2 - \omega^2)}(\cos \omega t - \cos nt)$$
where $n = (LC)^{-\frac{1}{2}}$ and E_m is the maximum e.m.f. Find the limiting value of i as $\omega \to n$.

4. In the integrand of
$$\int \frac{dx}{(1 + k \sin x)^{\frac{1}{2}}}$$
$-1 < k < 1$ and powers of k higher than the second can be ignored. Show, by expanding the integrand, that this indefinite integral is approximately equal to
$$(1 + \tfrac{3}{16}k^2)x + \tfrac{1}{2}k \cos x - \tfrac{3}{32}k^2 \sin 2x + \text{constant}$$

5. Write down the series expansion of
$$\log_e\left(\frac{1 + x}{1 - x}\right)$$
By substituting $x = 1/(2n + 1)$, prove that:
$$\log_e\left(\frac{n + 1}{n}\right) = 2\left[\frac{1}{2n + 1} + \frac{1}{3(2n + 1)^3} + \frac{1}{5(2n + 1)^5} + \cdots\right]$$
for all positive values of n.

Use this result to calculate $\log_e 2$ correct to five decimal places.

6. Find the value of $\int_0^{0.1} \log_e(1 + e^x)\, dx$ correct to four decimal places

7. Using the series for $\cosh x$, deduce that
$$\operatorname{sech}^2 x = 1 - x^2 + \tfrac{2}{3}x^4 + \cdots$$
and by integration derive a series for $\tanh x$, as far as the term in x^5.

8. By putting $x = \tfrac{1}{5}$ in the series for $(1 - x)^{-\frac{1}{2}}$, find the value of $\sqrt{5}$ to four decimal places.

9. By substituting in the series for $\operatorname{Tan}^{-1} x$, show that
$$\frac{\pi}{4} = 1 - \tfrac{1}{3} + \tfrac{1}{5} - \tfrac{1}{7} + \cdots$$

How many terms in this series do we need in order to calculate $\pi/4$ with an error not exceeding 0·0001 ? Show that

$$\text{Tan}^{-1} 1 = \text{Tan}^{-1} \tfrac{1}{2} + \text{Tan}^{-1} \tfrac{1}{3}$$

and deduce $\pi/4$ as the sum of two series. How many terms in these series do we need to calculate $\pi/4$ with an error not exceeding 0·0001 ?

10. A curve $y = f(x)$ touches the x-axis at the origin. Show that the first two terms in the Maclaurin's series of $f(x)$ are zero. Hence show that the radius of curvature at the origin is given by

$$\rho = \frac{1}{f''(0)} = \lim_{x \to 0} \left(\frac{x^2}{2y}\right)$$

and deduce a similar expression if the curve touches the y-axis at the origin.

Find the radius of curvature of the following, at the origin.

(i) $y = \dfrac{x^2(x-1)}{2-x}$ (ii) $4x^2 + y^2 - 8x = 0$

11. Evaluate $\int_0^{0\cdot 2} e^{-x^2} \tan x \, dx$ to four decimal places.

7
Leibniz' Rule

Illustrative Example 1

Find the third derivative of $y = e^{ax} \cdot \sinh bx$.

This function may be differentiated directly by using the product rule. If at each stage terms are collected together, we will find the following derivatives:

$Dy = e^{ax} \cdot b \cosh bx + ae^{ax} \cdot \sinh bx$
$D^2y = e^{ax} \cdot b^2 \sinh bx + 2ae^{ax} \cdot b \cosh bx + a^2 e^{ax} \cdot \sinh bx$
$D^3y = e^{ax} \cdot b^3 \cosh bx + 3ae^{ax} \cdot b^2 \sinh bx + 3a^2 e^{ax} \cdot b \cosh bx + a^3 e^{ax} \cdot \sinh bx$

Notice that this expression may be written:

$$D^3(u.v) = u \cdot D^3 v + 3 Du \cdot D^2 v + 3 D^2 u \cdot Dv + D^3 u \cdot v$$

where $u = e^{ax}$, $v = \sinh bx$. The right-hand side here bears a strong resemblance to the binomial expansion of $(a + b)^3$, which also has four terms with coefficients 1, 3, 3, 1. In fact, this is an example of a general rule for the nth derivative of a product $u.v$.

Leibniz' rule

$$D^n(u.v) = u \cdot D^n v + n Du \cdot D^{n-1} v + \frac{n(n-1)}{2!} D^2 u \cdot D^{n-2} v + \cdots$$
$$+ \frac{n(n-1) \ldots (n-r+1)}{r!} D^r u \cdot D^{n-r} v + \cdots$$
$$+ n D^{n-1} u \cdot Dv + D^n u \cdot v \quad (n \geqslant 0)$$

Notice that again the coefficients 1, n, $n(n-1)/2!, \ldots, n$, 1, are those which occur in the binomial expansion of $(a + b)^n$.

Illustrative Example 2

Find the nth derivative, $y^{(n)}(x)$, if $y(x) = (x^2 + 1) \cdot e^x$. Deduce $y^{(n)}(0)$ and hence derive the Maclaurin's series:

$$y(x) = 1 + x + \frac{3x^2}{2!} + \cdots$$

Using Leibniz' theorem with $u = (x^2 + 1)$ and $v = e^x$, we have:

$$y^{(n)}(x) = (x^2 + 1) D^n e^x + n D(x^2 + 1) \cdot D^{n-1} e^x$$
$$+ \frac{n(n-1)}{2!} D^2(x^2 + 1) \cdot D^{n-2} e^x + \cdots$$

In this case, since $D^3(x^2 + 1) = D^4(x^2 + 1) = \cdots = 0$, only the first three terms in $y^{(n)}(x)$ are non-zero, and we therefore have:

$$y^{(n)}(x) = (x^2 + 1) \cdot e^x + n \cdot 2x \cdot e^x + \tfrac{1}{2} n(n-1) \cdot 2 \cdot e^x$$
$$= e^x \{(x^2 + 1) + 2nx + n^2 - n\} \quad (n \geq 0)$$

To obtain $y^{(n)}(0)$, we simply put $x = 0$ into the above equation for $y^{(n)}(x)$:

$$y^{(n)}(0) = e^0 \{1 + n^2 - n\} = n^2 - n + 1$$

Therefore

$$y^{(1)}(0) = 1, \quad y^{(2)}(0) = 3, \quad y^{(3)}(0) = 7, \quad y^{(4)}(0) = 13, \ldots$$

Maclaurin's series for $y(x)$ may be written:

$$y(x) = y(0) + \frac{y^{(1)}(0)}{1!} x + \frac{y^{(2)}(0)}{2!} x^2 + \cdots$$

Substituting, we have:

$$y(x) = 1 + x + \frac{3x^2}{2!} + \frac{7x^3}{3!} + \cdots$$

Notice that the same series may be obtained by multiplication:

$$(1 + x^2)e^x = (1 + x^2)\left(1 + x + \frac{x^2}{2!} + \cdots\right)$$
$$= \left(1 + x + \frac{x^2}{2!} + \cdots\right) + \left(x^2 + x^3 + \frac{x^4}{2!} + \cdots\right)$$
$$= 1 + x + \frac{3x^2}{2!} + \cdots$$

Class Discussion Exercises 7

1. Find the sixth derivative $y^{(6)}(x)$ if $y = x^3 \sinh 2x$
2. If $y = x + \sqrt{(x^2 + 1)}$, show that:

$$\sqrt{(x^2 + 1)} y^{(1)} = y$$

and that:

$$(x^2 + 1)y^{(2)}(x) + xy^{(1)}(x) - y(x) = 0$$

Hence show that:

$$(x^2 + 1)y^{(n+2)}(x) + (2n + 1)xy^{(n+1)}(x) + (n^2 - 1)y^{(n)}(x) = 0$$
$$(n \geq 0)$$

3. Use the result of exercise 2 to show that
$$y^{(n+2)}(0) = -(n^2 - 1)y^{(n)}(0) \quad (n \geq 0)$$
and hence deduce the Maclaurin's series:
$$x + \sqrt{(x^2 + 1)} = 1 + \frac{x}{1!} + \frac{x^2}{2!} - \frac{3x^4}{4!} + \frac{45x^6}{6!} - \cdots$$

4. Check the series in exercise 3 by expanding $\sqrt{(1 + x^2)}$ as a binomial series.

5B. *Leibniz' Rule.* Assume that Leibniz' theorem is true for $n = k$, and by differentiation show that it is true for $n = k + 1$. Hence prove the theorem by induction.

Problems 7A

1. Evaluate:

   (i) $\dfrac{d^3}{dx^3}(x^2 e^{-x})$ (ii) $\dfrac{d^4}{dx^4}(x \cos^2 x)$ (iii) $\dfrac{d^3}{dx^3}(e^x \cos x)$

   (iv) $\dfrac{d^5}{dx^5}(x^2 \sin x)$ (v) $\dfrac{d^n}{dx^n}(x^2 e^x)$

2. When $y = e^{-x^2}$ find y' and y'' and prove that
$$y^{(n+1)}(x) + 2xy^{(n)}(x) + 2ny^{(n-1)}(x) = 0$$

3. Prove that when $y = \text{Tan}^{-1} x$
$$(1 + x^2)\frac{d^2y}{dx^2} + 2x\frac{dy}{dx} = 0$$
and hence show that
$$(1 + x^2)\frac{d^{n+2}y}{dx^2} + 2(n + 1)x\frac{d^{n+1}y}{dx^{n+1}} + n(n + 1)\frac{d^n y}{dx^n} = 0$$

4. When
$$y = \frac{\text{Sinh}^{-1} x}{\sqrt{1 + x^2}}$$
prove that

   (i) $(1 + x^2)y' + xy - 1 = 0$ (ii) $y^{(n+1)}(0) = -n^2 y^{(n-1)}(0)$

   Hence show that
$$\frac{\text{Sinh}^{-1} x}{\sqrt{1 + x^2}} = x - \tfrac{2}{3}x^3 + \tfrac{8}{15}x^5 - \tfrac{16}{35}x^7 + \cdots$$

Problems 7B

1. When $y = \cosh(k \operatorname{Sin}^{-1} x)$ prove that
$$(1 - x^2)y'' - xy' - k^2 y = 0 \quad \text{and} \quad y^{(n+2)}(0) = (n^2 + k^2)y^{(n)}(0)$$
Hence deduce that
$$\cosh(k \operatorname{Sin}^{-1} x) = 1 + \frac{k^2}{2!}x^2 + \frac{k^2(k^2 + 2^2)}{4!}x^4$$
$$+ \frac{k^2(k^2 + 2^2)(k^2 + 4^2)}{6!}x^6 + \cdots$$

2. When $y = \sin(a \operatorname{Sin}^{-1} x)$, show that
$$(1 - x^2)y'' - xy' + a^2 y = 0 \quad \text{and} \quad y^{(n+2)}(0) = (n^2 - a^2)y^{(n)}(0)$$
Hence obtain the first three terms in the Maclaurin expansion of $\sin(a \operatorname{Sin}^{-1} x)$. By writing $\operatorname{Sin}^{-1} x = \theta$, prove that
$$\sin 2\theta = 2 \sin \theta - \sin^3 \theta - \tfrac{1}{4}\sin^5 \theta + \cdots$$

3. When $y = (\operatorname{Sin}^{-1} x)^2$, show that $(1 - x^2)y'' - xy' = 2$ and hence find the first four terms in the series expansion of $(\operatorname{Sin}^{-1} x)^2$.

4. When $y = \{x + \sqrt{(x^2 + 1)}\}^r$, show that
$$(x^2 + 1)y'' + xy' - r^2 y = 0$$
Hence prove that $y^{(n+2)}(0) = (r^2 - n^2)y^{(n)}(0)$.

8
Integration using Trigonometric and Hyperbolic Functions

There are two main techniques which are useful with the type of integral which we discuss here, where the integrand contains trigonometric functions. The first depends upon using differential identities such as $\cos x\, dx = d(\sin x)$, or $\sec^2 x\, dx = d(\tan x)$, which if spotted in the integrand can often lead to a simple evaluation. The second technique is to try using the standard trigonometric identities, and for this purpose we give below a table of the most common identities in the form they are most often needed. The corresponding hyperbolic identities may be derived if necessary from this table by means of Osborne's Rule.

$$\sin^2 \theta = \tfrac{1}{2}(1 - \cos 2\theta)$$
$$\cos^2 \theta = \tfrac{1}{2}(1 + \cos 2\theta)$$
$$\sin m\theta \cos n\theta = \tfrac{1}{2} \sin(m+n)\theta + \tfrac{1}{2} \sin(m-n)\theta$$
$$\cos m\theta \cos n\theta = \tfrac{1}{2} \cos(m+n)\theta + \tfrac{1}{2} \cos(m-n)\theta$$
$$\sin m\theta \sin n\theta = -\tfrac{1}{2} \cos(m+n)\theta + \tfrac{1}{2} \cos(m-n)\theta$$

The substitution $t = \tan(\tfrac{1}{2}\theta)$ implies:

(i) $\cos \theta = \dfrac{1 - t^2}{1 + t^2}$ \quad (ii) $\sin \theta = \dfrac{2t}{1 + t^2}$

(iii) $d\theta = \dfrac{2\,dt}{1 + t^2}$

Illustrative Example 1

Evaluate:

(i) $\displaystyle\int \sin^4 x \cos x\, dx,$ \quad (ii) $\displaystyle\int \sin^5 x\, dx,$ \quad (iii) $\displaystyle\int \sin^4 x\, dx$

(i) Noticing that $\cos x\, dx = d(\sin x)$, we write:

$$\int \sin^4 x \cos x\, dx = \int \sin^4 x\, d(\sin x) = \tfrac{1}{5} \sin^5 x + c$$

(ii) Here, we may convert the integral into one with respect to $\cos x$, using $d(\cos x) = \sin x\, dx$:

$$\int \sin^5 x\, dx = -\int \sin^4 x\, d(\cos x)$$
$$= -\int (1 - \cos^2 x)^2 d(\cos x)$$
$$= -\int 1 - 2\cos^2 x + \cos^4 x\, d(\cos x)$$
$$= -\cos x + \tfrac{2}{3}\cos^3 x - \tfrac{1}{5}\cos^5 x + c$$

(iii) This is the only one of these three problems where we cannot make use of differential notation directly; in this case we use the first trigonometrical identity to reduce the power of $\sin x$ in the integrand:

$$\sin^4 x = (\sin^2 x)^2 = \tfrac{1}{4}(1 - \cos 2x)^2 = \tfrac{1}{4}(1 - 2\cos 2x + \cos^2 2x)$$
$$= \tfrac{1}{4}(1 - 2\cos 2x + \tfrac{1}{2}(1 + \cos 4x))$$
$$= \tfrac{3}{8} - \tfrac{1}{2}\cos 2x + \tfrac{1}{8}\cos 4x$$

Hence, on integrating

$$\int \sin^4 x\, dx = \tfrac{3}{8}x - \tfrac{1}{4}\sin 2x + \tfrac{1}{32}\sin 4x + c$$

TRIGONOMETRIC AND HYPERBOLIC SUBSTITUTIONS

One of the most useful types of substitution in formal integration involves the use of trigonometric or hyperbolic functions. Whenever the integrand contains an expression which may be reduced conveniently as a trigonometric or hyperbolic identity, a substitution of this type is worth trying. For instance, the expression $\sqrt{(4 - x^2)}$ occurs in example 2, which reduces to $2\cos\theta$ on substitution of $x = 2\sin\theta$, thus simplifying the integral. Other expressions in the integrand suggest other substitutions, and a table of common substitutions which form the basis of the work in this section is given below. There is no great need to memorise this table, since each of the substitutions is suggested by the well-known identities.

Illustrative Example 2

Integrate:

$$I = \int \frac{dx}{\sqrt{(4 - x^2)}}$$

Although this is one of the standard integrals, we shall treat it here as an example of the general method of trigonometrical substitution.

The expression $\sqrt{(4 - x^2)}$ suggests this, since putting $x = 2 \sin \theta$, we have:

$$2\sqrt{(1 - \sin^2 \theta)} = 2 \cos \theta$$

Differentiating $x = 2 \sin \theta$ gives $dx = 2 \cos \theta d\theta$,
Therefore

$$I = \int \frac{1}{2 \cos \theta} 2 \cos \theta \, d\theta = \int d\theta = \theta + c$$

Hence,

$$I = \mathrm{Sin}^{-1}\left(\frac{x}{2}\right) + c$$

In this example the expression $\sqrt{(a^2 - x^2)}$ suggested the sine substitution because of the identity $1 - \sin^2 \theta = \cos^2 \theta$. An expression such as $\sqrt{(a^2 + x^2)}$ might suggest either $x = a \tan \theta$ (because: $1 + \tan^2 \theta = \sec^2 \theta$) or $x = a \sinh \theta$ (because $1 + \sinh^2 \theta = \cosh^2 \theta$). We give in the table below both trigonometrical substitutions for certain expressions, and hyperbolic alternatives in each case:

$(a^2 - x^2)$	$x = a \sin \theta$	$x = a \tanh \theta$
$(x^2 - a^2)$	$x = a \sec \theta$	$x = a \cosh \theta$
$(x^2 + a^2)$	$x = a \tan \theta$	$x = a \sinh \theta$

Illustrative Example 3

Integrate:

$$I = \int \frac{dx}{x^2(x^2 - 4)^{\frac{1}{2}}}$$

Using $x = 2 \cosh \theta$, $dx = 2 \sinh \theta \, d\theta$,

$$I = \int \frac{2 \sinh \theta}{8 \cosh^2 \theta \sinh \theta} d\theta$$

$$= \frac{1}{4} \int \mathrm{sech}^2 \theta \, d\theta$$

$$= \tfrac{1}{4} \tanh \theta + c$$

If we now wish to express this in terms of x, we may use

$$\cosh \theta = \tfrac{1}{2}x, \quad \sinh \theta = \sqrt{(\cosh^2 \theta - 1)} = \tfrac{1}{2}\sqrt{(x^2 - 4)}$$

Therefore

$$I = \frac{(x^2 - 4)^{\frac{1}{2}}}{4x} + c$$

Class Discussion Exercises 8

1. Integrate:

   (i) $\displaystyle\int \sqrt{(1 + \cos x)}\, dx$
   (ii) $\displaystyle\int \cos^5 t \sin t\, dt$
   (iii) $\displaystyle\int \frac{\cos\theta\, d\theta}{\sin^3\theta}$
   (iv) $\displaystyle\int_{-\pi}^{\pi} \cos 2u \cos 5u\, du$
   (v) $\displaystyle\int \tan^2 x\, dx$
   (vi) $\displaystyle\int \sec^2\theta \tan\theta\, d\theta$

2. Use Osborne's rule to deduce the hyperbolic forms of the identities in the first part of the table given above. Integrate:

   (i) $\displaystyle\int \cosh^2 x \sinh x\, dx$
   (ii) $\displaystyle\int \sinh^4 t\, dt$
   (iii) $\displaystyle\int \sqrt{(1 + \cosh x)}\, dx$

3. Show that if $t = \tan(\tfrac{1}{2}\theta)$, then $\tan\theta = 2t/(1 + t^2)$. By considering a right-angled triangle containing the angle θ, or otherwise, obtain the final set of formulae in the above table result. Integrate:

   $$\int \frac{d\theta}{\cos\theta + 2\sin\theta + 3}$$

4. Show that each of the expressions given in the table reduces conveniently under the substitution given.

5. Integrate the integral in illustrative example 3 using the trigonometric substitution.

6. Evaluate:

   $$\int_{2/\sqrt{3}}^{\infty} \frac{dx}{x^2\sqrt{(x^2 + 4)}}$$

7B. By completing the square, evaluate:

   $$\int_{-1}^{0} \sqrt{(3 - 2x - x^2)}\, dx$$

8B. The standard hyperbolic integrals given in the following table are similar to the standard trigonometric integrals. In fact they differ only by the occasional sign change and the sign may be determined by differentiation of the function on the right. In the following table where the signs have been omitted on the right, fill in the appropriate signs. (The purpose of this exercise is to show that the table need *not* be memorised.)

$\int \sinh x \, dx$	$\cosh x$
$\int \cosh x \, dx$	$\sinh x$
$\int \tanh x \, dx$	$\log_e |\cosh x|$
$\int \coth x \, dx$	$\log_e |\sinh x|$
$\int \operatorname{sech}^2 x \, dx$	$\tanh x$
$\int \operatorname{cosech}^2 x \, dx$	$\coth x$
$\int \operatorname{sech} x \tanh x \, dx$	$\operatorname{sech} x$
$\int \operatorname{cosech} x \, dx$	$\log |\tanh \tfrac{1}{2}x|$
$\int \operatorname{sech} x \, dx$	$\log |\operatorname{sech} x \quad \tanh x|$*

*(Note *where* the sign change occurs in this integral.)

Problems 8A

Integrate:

1. (i) $\displaystyle\int \sin^3 t \, dt$ (ii) $\displaystyle\int_0^{\pi/2} \sin^2 \theta \cos^3 \theta \, d\theta$

   (iii) $\displaystyle\int \frac{\cos 3x}{1 + \sin 3x} \, dx$ (iv) $\displaystyle\int \sin^2 x \cos^2 x \, dx$

   (v) $\displaystyle\int \tanh^2 x \, dx$ (vi) $\displaystyle\int \coth x \, dx$

2. (i) $\displaystyle\int \tan^3 \theta \, d\theta$ (ii) $\displaystyle\int \sqrt{(1 + \cos x)} \, dx$

   (iii) $\displaystyle\int \sin 5x \cos 7x \, dx$ (iv) $\displaystyle\int \cos x \sqrt{(\sin x)} \, dx$

   (v) $\displaystyle\int \cosh^5 \theta \, d\theta$ (vi) $\displaystyle\int \sqrt{(\cosh x - 1)} \, dx$

3. (i) $\displaystyle\int \cosh^2 u \, du$ (ii) $\displaystyle\int \sinh^2 u \, du$

   (iii) $\displaystyle\int \sinh^5 t \cosh t \, dt$ (iv) $\displaystyle\int \sqrt{(\cosh x - 1)} \, dx$

4. Use the substitution $t = \tan(\theta/2)$ to evaluate:

   (i) $\displaystyle\int \frac{d\theta}{3 + 5\cos\theta}$ (ii) $\displaystyle\int \operatorname{cosec} \theta \, d\theta$

   (iii) $\displaystyle\int \frac{d\theta}{1 + \sin\theta}$ (iv) $\displaystyle\int_{\pi/3}^{\pi/2} \frac{d\theta}{2\sin\theta + \cos\theta - 1}$

   (v) $\displaystyle\int_0^{\pi/2} \frac{d\theta}{2 - \cos\theta}$

52 Integration using Trigonometric and Hyperbolic Functions

5. Find (a) the mean value, (b) the root-mean-square value of the following functions over the interval indicated:

   (i) $i = i_1 \cos \omega t$; $0 \leqslant t \leqslant \dfrac{2\pi}{\omega}$

   (ii) $i = \cos t + \sin t$; $0 \leqslant t \leqslant \pi$

   (iii) $i = i_1 \sin(\omega t + \phi_1) + i_2 \sin(\omega t + \phi_2)$; $0 \leqslant t \leqslant \dfrac{2\pi}{\omega}$

   (iv) $i = i_0 + i_1 \sin(\omega t + \phi)$; $0 \leqslant t \leqslant \dfrac{2\pi}{\omega}$

6. Integrate:

   (i) $\displaystyle\int_{a/2}^{a} \dfrac{dx}{x(a^2 - x^2)^{\frac{1}{2}}}$

   (ii) $\displaystyle\int_{0}^{2} \dfrac{dx}{(4 + x^2)^2}$

   (iii) $\displaystyle\int_{0}^{1} \dfrac{x^2}{(x^2 + 1)^{\frac{3}{2}}} dx$

   (iv) $\displaystyle\int_{0}^{1} x\sqrt{(1 + x^2)}\, dx$

   (v) $\displaystyle\int_{0}^{a} \sqrt{(a^2 - x^2)}\, dx$

   (vi) $\displaystyle\int_{0}^{1} \sqrt{(x^2 + 1)}\, dx$

7. (i) $\displaystyle\int (x^2 - 9)^{\frac{1}{2}} dx$

   (ii) $\displaystyle\int_{0}^{3} \dfrac{du}{(u^2 + 4)^{\frac{3}{2}}}$

   (iii) $\displaystyle\int \dfrac{dx}{x^2\sqrt{(4 - x^2)}}$

   (iv) $\displaystyle\int_{3}^{5} \dfrac{\sqrt{(u^2 - 9)}}{u}\, du$

   (v) $\displaystyle\int \dfrac{dx}{(1 - 4x^2)^{\frac{3}{2}}}$

   (vi) $\displaystyle\int \sqrt{(9x^2 + 1)}\, dx$

8. Use a hyperbolic substitution to evaluate:

$$\int_{0}^{x} \dfrac{dt}{\sqrt{(t^2 - a^2)}}$$

By comparison with the table of standard integrals, show that:

$$\cosh^{-1}\left(\dfrac{x}{a}\right) = \log_e \left|\dfrac{x + \sqrt{(x^2 - a^2)}}{a}\right|$$

9. Integrate:

$$\int \dfrac{dt}{a^2 - t^2}$$

(i) by the substitution $t = a \tanh \theta$, (ii) by partial fractions. Show that:

$$\tfrac{1}{2} \log \left|\dfrac{a + x}{a - x}\right| = \tanh^{-1}\left(\dfrac{x}{a}\right)$$

10. Show by integrating $\int_{0}^{a} dt/\sqrt{(t^2 + a^2)}$ that:

$$\log \left|\dfrac{x + \sqrt{(x^2 + a^2)}}{a}\right| = \sinh^{-1}\left(\dfrac{x}{a}\right)$$

11. A probability density function for the random variable x is
$$p(x) = \begin{cases} k(a^2 - x^2)^{\frac{1}{2}}: & 0 \leq x \leq a \\ 0: & \text{otherwise} \end{cases}$$
Given that $\int_{-\infty}^{\infty} p(x)\, dx = 1$, find k. Hence find:
(i) the mean value of x:
$$\mu = E[x] = \int_{-\infty}^{\infty} x p(x)\, dx$$
(ii) the variance of x:
$$\sigma^2 = E[(x - \mu)^2] = \int_{-\infty}^{\infty} (x - \mu)^2 p(x)\, dx$$

Problems 8B

1. Find:

(i) $\displaystyle\int \frac{dx}{\cos x(1 + \cos x)}$ (ii) $\displaystyle\int \frac{dx}{\sin x(1 + \sin x)}$

2. (i) $\displaystyle\int \frac{d\theta}{\sqrt{(1 - \cos 2\theta)}}$ (ii) $\displaystyle\int \frac{\sin 2x\, dx}{1 + \cos^2 x}$

(iii) $\displaystyle\int \tan^5 x\, dx$ (iv) $\displaystyle\int \tanh^3 \theta\, d\theta$

(v) $\displaystyle\int \tanh \theta\, \text{sech}^2 \theta\, d\theta$ (vi) $\displaystyle\int \frac{d\theta}{\sqrt{(\cosh \theta - 1)}}$

3. Find the total length of the curve: $x = a\cos^3 t$, $y = a\sin^3 t$.

4. Find the length of the curve $y = \log(\sec x)$ from $x = 0$ to $x = \pi/4$.

5. Find the volume of the solid obtained by rotating about Ox the area enclosed between the x-axis and the curve:
$$y = a\sin \alpha x + b\sin 2\alpha x, \quad 0 \leq x \leq \pi/2\alpha$$

6. Show that:

(i) $\displaystyle\int_{-\pi}^{\pi} \sin mx \sin nx\, dx = \begin{cases} 0: & m \neq n \\ \pi: & m = n \end{cases}$

(ii) $\displaystyle\int_{-\pi}^{\pi} \sin mx \cos nx\, dx = 0$

(iii) $\displaystyle\int_{-\pi}^{\pi} \cos mx \cos nx\, dx = \begin{cases} 0: & m \neq n \\ \pi: & m = n \end{cases}$

where $m, n \in Z^+$.

7. By completing the square, integrate:

   (i) $\int_1^2 (x^2 + 2x + 2)^{\frac{1}{2}} \, dx$ (ii) $\int_1^{\frac{3}{2}} (4t^2 - 4t)^{\frac{1}{2}} \, dt$

8. Integrate:

   (i) $\int \dfrac{du}{(1 + u^2)^2}$ (ii) $\int_0^\infty \dfrac{dx}{(x^2 + 1) + x(x^2 + 1)^{\frac{1}{2}}}$

   (iii) $\int_2^\infty \dfrac{dx}{x(x^2 + 4)^{\frac{1}{2}}}$

9. $\int_0^2 \left(\dfrac{4 - x}{4 + x}\right)^{\frac{1}{2}} dx$

10. $\int_1^{\sqrt{2}} \dfrac{dx}{x(x^4 - 1)^{\frac{1}{2}}}$

11. $\int_0^1 \sqrt{(x^2 + x + 1)} \, dx$

9
Reduction Formulae

In some cases when we need to evaluate an integral involving an integer n we can find it in terms of a similar integral but with the number n reduced. For instance in the example below, application of the parts formula allows us to find $\int \sin^n x \, dx$ in terms of $\int \sin^{n-2} x \, dx$. A relation such as this is known as a *reduction formula*, since it reduces the value of n in the integral we want to find. By successive application of the formula we may reduce integrals with high values of n to those with low values, which can then be evaluated more easily.

Much of the writing involved in applying reduction formulae may be saved by writing I_n to represent the integral, e.g. $I_n = \int \sin^n x \, dx$.

Illustrative Example 1

Show that

$$\int_0^{\pi/2} \sin^n x \, dx = \frac{(n-1)}{n} \int_0^{\pi/2} \sin^{n-2} x \, dx \quad \text{for } n \geq 2$$

Hence evaluate:

$$\int_0^{\pi/2} \sin^7 x \, dx$$

Integrating by parts,

$$I_n = \int_0^{\pi/2} \sin^n x \, dx = \int_0^{\pi/2} \sin^{n-1} x \, d(-\cos x)$$

$$= [-\sin^{n-1} x \cos x]_0^{\pi/2} + \int_0^{\pi/2} \cos^2 x (n-1) \sin^{n-2} x \, dx$$

$$= 0 + (n-1) \int_0^{\pi/2} \sin^{n-2} x (1 - \sin^2 x) dx$$

$$= (n-1) \int_0^{\pi/2} \sin^{n-2} x \, dx - (n-1) \int_0^{\pi/2} \sin^n x \, dx$$

Therefore
$$I_n = (n-1)I_{n-2} - (n-1)I_n$$
Solving this equation for I_n gives the reduction formula:
$$I_n = \frac{(n-1)}{n} I_{n-2}$$
i.e.
$$\int_0^{\pi/2} \sin^n x \, dx = \frac{(n-1)}{n} \int_0^{\pi/2} \sin^{n-2} x \, dx$$

To find I_7, we first put $n = 7$:
$$I_7 = \tfrac{6}{7} I_5$$
and so reduce the problem to that of finding I_5. But now we may use the formula again:
$$I_5 = \tfrac{4}{5} I_3$$
and
$$I_3 = \tfrac{2}{3} I_1 = \frac{2}{3} \int_0^{\pi/2} \sin x \, dx = \tfrac{2}{3} \cdot (1)$$

Therefore
$$I_7 = \tfrac{6}{7} \cdot \tfrac{4}{5} \cdot \tfrac{2}{3} \cdot (1) = \tfrac{16}{35}$$

Class Discussion Exercises 9

1. If $I_n = \int_0^\infty x^n e^{-x} \, dx$, prove that $I_n = n I_{n-1}$, $n \geqslant 1$. Hence evaluate I_6.
2. Evaluate $\int_0^{\pi/2} \sin^6 x \, dx$, using the reduction formula proved in example 1. Show that in general:

$$\int_0^{\pi/2} \sin^n x \, dx = \begin{cases} \dfrac{(n-1)(n-3)\ldots 3.1}{n(n-2)\ldots 4.2} \cdot \dfrac{\pi}{2} & (n \text{ even}) \\ \dfrac{(n-1)(n-3)\ldots 4.2}{n(n-2)\ldots 5.3} \cdot 1 & (n \text{ odd}) \end{cases}$$

This is one of *Wallis' formulae*. The same formula holds for $\int_0^{\pi/2} \cos^n x \, dx$.

3. If $I_n = \int \tan^n x \, dx$, show that
$$I_n = \frac{\tan^{n-1} x}{n-1} - I_{n-2}, \quad n \geqslant 2$$
Hence evaluate $\int \tan^6 x \, dx$.

4. Find a reduction formula for $I_m = \int_0^a (a^2 - x^2)^m \, dx$. Hence evaluate $\int_0^2 (4 - x^2)^{\frac{5}{2}} \, dx$.

Problems 9A

1. Show that $\int_0^{\pi/2} \sin^n x \, dx = \int_0^{\pi/2} \cos^n x \, dx$ by using the substitution $x = \pi/2 - x'$. Using Wallis' formula evaluate:

   (i) $\displaystyle\int_0^{\pi/2} \cos^4 x \, dx$

   (ii) $\displaystyle\int_0^{\pi/2} \cos^7 x \, dx$

2. If $I_m = \int_0^\pi e^{-x} \sin^m x \, dx$, show that
$$(m^2 + 1)I_m = m(m - 1)I_{m-2}, \quad \text{for } m \geq 2$$
Hence evaluate $\int_0^\pi e^{-x} \sin^4 x \, dx$.

3. If $I_n = \int \cot^n x \, dx$, show that
$$I_n = \frac{-\cot^{n-1} x}{n - 1} - I_{n-2} \quad (n \geq 2)$$
hence find

   (i) $\displaystyle\int \cot^6 x \, dx$

   (ii) $\displaystyle\int \cot^7 x \, dx$

4. Obtain a reduction formula for $\int x^n e^{-x} \, dx$. Hence find $\int_0^1 x^5 e^{-x} \, dx$, showing that your answer is positive and less than 1.

5. Obtain a reduction formula for $\int_0^{\pi/2} x \sin^n x \, dx$. Hence evaluate $\int_0^{\pi/2} x \sin^4 x \, dx$.

6. If $I_n = \int_0^{\pi/2} x^n \sin x \, dx$, show that
$$I_n = n\left(\frac{\pi}{2}\right)^{n-1} - n(n - 1)I_{n-2}, \quad \text{for } n \geq 2$$

7. Show that $u_n = \int_0^a x^n (a - x)^{\frac{1}{2}} \, dx$ satisfies the reduction formula
$$u_n = \frac{2na}{2n + 3} u_{n-1}, \quad \text{for } n \geq 1$$
and calculate u_3.

Problems 9B

1. If $I_n = \displaystyle\int \frac{x^n}{1 + x^2} \, dx$, show that
$$I_n = \frac{x^{n-1}}{n - 1} - I_{n-2} \quad (n \geq 2)$$

Hence find

(i) $\displaystyle\int_0^1 \frac{x^7}{1+x^2}\,dx$ (ii) $\displaystyle\int \frac{x^6}{1+x^2}\,dx$

2. Find a reduction formula for $\int x^n(2x+c)^{-\frac{1}{2}}\,dx$ and hence evaluate $\int_0^1 x^3(2x+1)^{-\frac{1}{2}}\,dx$.

3. If $I_{m,n} = \int_0^{\pi/2} \cos^m x \sin^n x\,dx$, show that
$$I_{m,n} = \frac{(n-1)}{(m+n)} I_{m,n-2} \quad (n \geqslant 2)$$
and similarly
$$I_{m,n} = \frac{(m-1)}{(m+n)} I_{m-2,n} \quad (m \geqslant 2)$$
Hence evaluate:

(i) $\displaystyle\int_0^{\pi/2} \cos^3 x \sin^2 x\,dx$ (ii) $\displaystyle\int_0^{\pi/2} \cos^4 x \sin^2 x\,dx$

(iii) $\displaystyle\int_0^{\pi/2} \sin^3 x \cos^3 x\,dx$

4. If $I_m = \displaystyle\int \frac{dt}{(1+t)t^m}$, show that
$$I_m = \frac{-1}{(m-1)t^{m-1}} - I_{m-1}, \quad \text{for } m > 1$$

5. If $I_n = \int_0^1 (1-x^3)^n\,dx$, show that:
$$(3n+1)I_n = 3nI_{n-1}, \quad (n \geqslant 2)$$
and find I_3.

6. Find a reduction formula for the integral:
$$I_n = \int (\log_e x)^n\,dx$$

10
Polar Coordinates

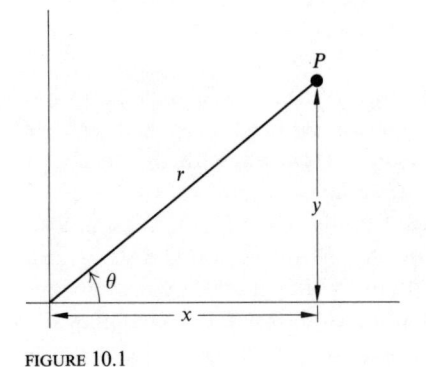

The curves that we have been studying so far have all been expressed in terms of x and y coordinates (rectangular cartesian coordinates). For some physical situations it is much easier to use *polar* coordinates (r,θ), since the curves involved have simple polar equations.

In the same way that a point P (figure 10.1) is fixed when we know the cartesian coordinates (x, y), it is also fixed when we know the polar coordinates (r, θ), where r is the distance of P from the origin and θ is the angle between OP and Ox, measured positive anticlockwise. From figure 10.1 we have the following relations between cartesians and polars, which allow us to make a change between one set of coordinates and the other:

FIGURE 10.1

$$x = r \cos \theta \qquad r^2 = x^2 + y^2$$
$$y = r \sin \theta \qquad \tan \theta = \frac{y}{x}$$

Illustrative Example 1

Plot the points with the following polar coordinates (r, θ): $P_1(2, \tfrac{1}{3}\pi)$, $P_2(2, -\tfrac{1}{2}\pi)$, $P_3(2, 3\pi/2)$, $P_4(1, 0)$, $P_5(-1, \tfrac{1}{4}\pi)$, $P_6(0, 3\cdot7)$.

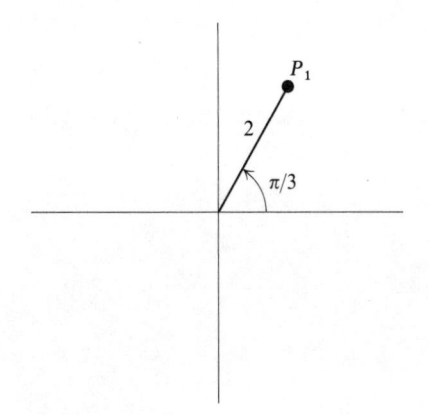

FIGURE 10.2

FIGURE 10.3

60 Polar Coordinates

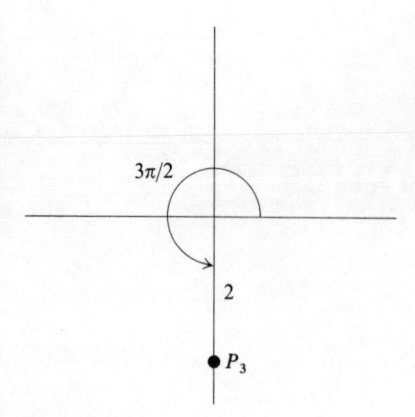

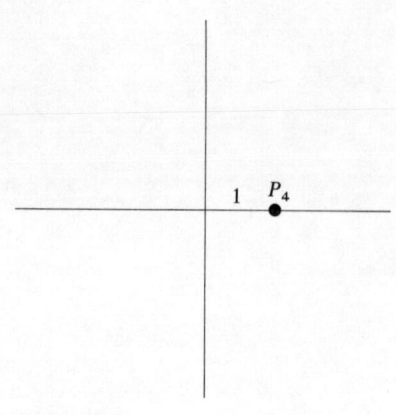

FIGURE 10.4

FIGURE 10.5

Points P_1 to P_4 are plotted in figures 10.2–5 and their polar coordinates are indicated on the diagrams. Notice that P_2 has negative $\theta = -\frac{1}{2}\pi$, and so we must measure a right angle in the *clockwise* direction. Here, P_2 and P_3 are the same point, so that we have two sets of polar coordinates representing the same point. We can also give other polar coordinates for this point, since $(2, 3\pi/2 + 2\pi)$, $(2, 3\pi/2 + 4\pi)$, ... all give the point P_2. In general, we may always add or subtract multiples of 2π to or from θ without changing the point.

In figure 10.6 we have plotted P_5 where $r = -1$. Using the equations $x = r \cos \theta$, $y = r \sin \theta$, we get for P_5: $x = -\frac{1}{2}$, $y = -\frac{1}{2}$, and so P_5 is in the third quadrant. We must interpret $r = -1$, therefore, to mean that we move a distance 1 in the *opposite direction* to the line making $\frac{1}{4}\pi$ with Ox; this indicates the general way in which we interpret negative r.

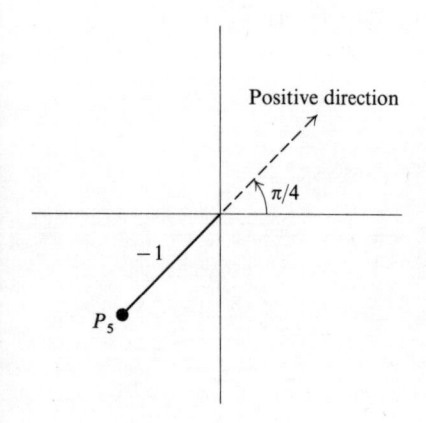

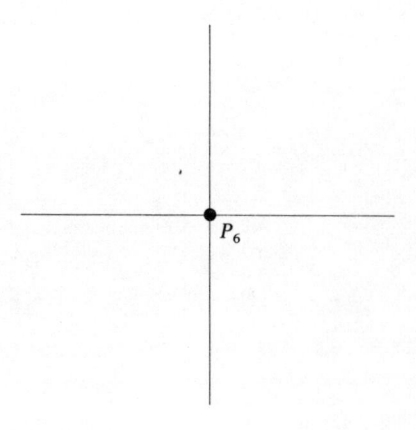

FIGURE 10.6

FIGURE 10.7

Illustrative Example 2

Sketch the curve

$$r = \frac{1}{2 + \cos\theta}$$

This curve can be sketched simply by calculating r for various values of θ and plotting a selection of points (figure 10.8). This method always works, but we can sometimes shorten the working by the following considerations:

Equation	Symmetry
(i) r = even function of θ	symmetric about x-axis
(ii) r = odd function of θ	symmetric about y-axis
(iii) only even functions of r present	symmetric about origin

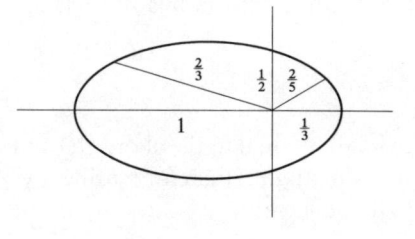

FIGURE 10.8 *The ellipse* $r = \dfrac{1}{2 + \cos\theta}$

In this example r is an even function of θ since $\cos\theta$ is even. Therefore the curve is symmetric about the x-axis.

Plotting the following points together with the symmetry about Ox, gives us the curve shown in figure 10.8, an ellipse with focus at the origin.

$$\theta = 0, \qquad r = \tfrac{1}{3}$$
$$\theta = \frac{\pi}{6}, \qquad r = \tfrac{2}{5}$$
$$\theta = \tfrac{1}{2}\pi, \qquad r = \tfrac{1}{2}$$
$$\theta = \frac{\pi}{6}, \qquad r = \tfrac{2}{3}$$
$$\theta = \pi, \qquad r = 1$$

Illustrative Example 3: Area

Find the area A of the loop of $r = 2a\cos\theta$ (figure 10.9).

Let the area be divided into fine sectors of angle $\delta\theta$. The area of each sector is approximately the area of a sector of a circle with radius r and angle $\delta\theta$. Therefore

$$\text{Area of sector} \approx \tfrac{1}{2}r^2\,\delta\theta$$

Summing over sectors and letting $\delta\theta \to 0$,

$$A = \int_{-\pi/2}^{\pi/2} \tfrac{1}{2}r^2\,d\theta = \int_{-\pi/2}^{\pi/2} \tfrac{1}{2}\cdot 4a^2\cos^2\theta\,d\theta = \pi a^2$$

The formula for area has been derived quite generally in this example, so that we have for the area between limits $\theta = \alpha$, $\theta = \beta$:

$$\boxed{\text{Area} = \int_\alpha^\beta \tfrac{1}{2}r^2\,d\theta}$$

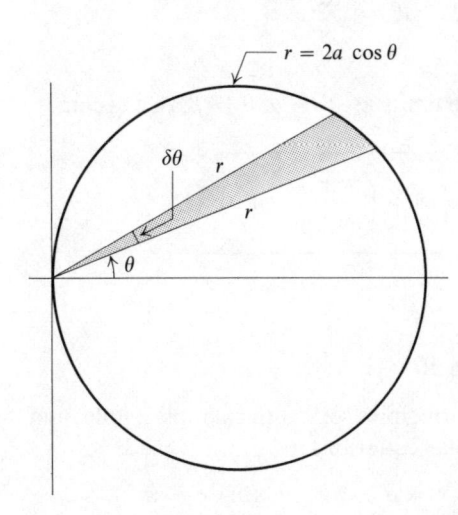

FIGURE 10.9

Illustrative Example 4: Arc Length

Find the length of the spiral $r = e^\theta$ (figure 10.10), from $\theta = 0$ to $\theta = \frac{1}{4}\pi$.

Let the arc be divided into elements δs, subtending an angle $\delta \theta$ at O. For the element PQ, $OP = r$ and $OQ = r + \delta r$. Hence RP is approximately equal to the arc of a circle with radius r, angle $\delta\theta$. Therefore

$$RP \approx r\delta\theta$$

If δs is small, the arc PQ is approximately equal to the chord PQ and angle PQR is approximately a right angle; therefore using Pythagoras' theorem in triangle PQR, we have:

$$\delta s^2 \approx r^2 \, \delta\theta^2 + \delta r^2$$

Therefore

$$\delta s \approx \left\{ r^2 + \left(\frac{\delta r}{\delta \theta}\right)^2 \right\}^{\frac{1}{2}} \delta\theta$$

Letting $\delta\theta \to 0$ and summing, we have for the arc length, s:

$$s = \int_{\theta=0}^{\theta=\pi/4} ds = \int_0^{\pi/4} \left\{ r^2 + \left(\frac{dr}{d\theta}\right)^2 \right\}^{\frac{1}{2}} d\theta$$

Putting $r = e^\theta$, we have

$$s = \int_0^{\pi/4} \sqrt{2} \, e^\theta \, d\theta = \sqrt{2}(e^{\frac{1}{4}\pi} - 1)$$

In the general case, where the limits are $\theta = \alpha$, $\theta = \beta$, this becomes:

$$\boxed{s = \int_\alpha^\beta \left\{ r^2 + \left(\frac{dr}{d\theta}\right)^2 \right\}^{\frac{1}{2}} d\theta}$$

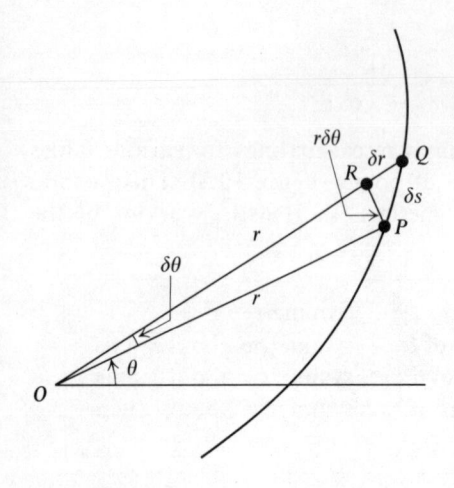

FIGURE 10.10

Class Discussion Exercises 10

1. Show that the symmetry principles set out in example 2 hold, and sketch the curves with polar equations:

    (i) $r \cos \theta = 1$ (ii) $r = a$ (iii) $r = a\theta$
    (iv) $\theta = \pi/6$ (v) $r = 2a \sin \theta$ (vi) $r = a \sin 2\theta$
    (vii) $r^2 = a^2 \sin \theta$ (viii) $r = 1 + \cos \theta$

2. Find by geometrical reasoning the polar equation of the circle with radius a shown in figure 10.11. Find the cartesian equation of this circle.
3. Find the area of one loop of the curve given by $r = a \sin 2\theta$.
4. Find the area inside the circle $r = 2 \sin \theta$ and outside $r = 1$.
5. Find the length of the arc of

$$r = \sin \theta + 2 \cos \theta$$

from $\theta = 0$ to $\theta = \pi$. Transform to cartesians and show that the curve is a circle with centre $(1, \tfrac{1}{2})$.

6. Show from triangle PQR in figure 10.12 that

$$\tan \angle QPR \approx r \frac{\delta \theta}{\delta r}$$

Deduce that the angle ϕ between the radius vector OP and the tangent is given by:

$$\boxed{\tan \phi = r \frac{d\theta}{dr}}$$

Find $\tan \phi$ for the curve $r = a(1 + \cos \theta)$ at $\theta = 0, \tfrac{1}{2}\pi, \pi$. Sketch the curve, using this information.

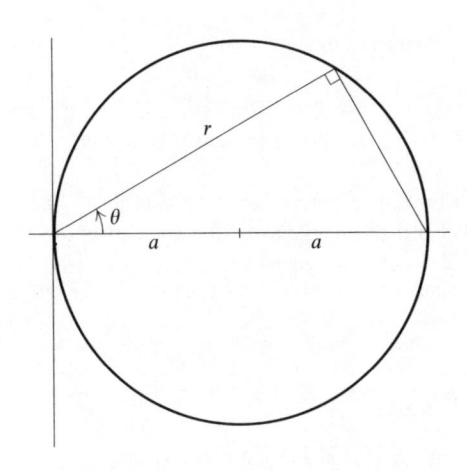

FIGURE 10.11

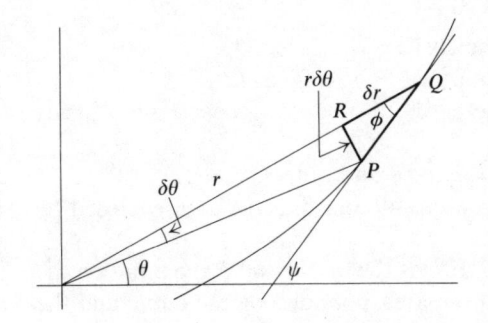

FIGURE 10.12

Problems 10A

1. Transform the following equations into rectangular cartesians and identify the curves:
   - (i) $r = a$
   - (ii) $r = \cos \theta + \sin \theta$
   - (iii) $\theta = 3\pi/4$
   - (iv) $r \sin \theta = 1$
   - (v) $r = 2a \sin \theta$
   - (vi) $\theta = -\tfrac{1}{2}\pi$

2. Transform into polars:
   - (i) $x^2 + y^2 - 2ay = 0$
   - (ii) $x - y = 1$
   - (iii) $x^3 + xy^2 - ay = 0$
   - (iv) $y = x \tan \alpha$

3. Sketch the curves:
   - (i) $r = e^\theta$
   - (ii) $r = a(1 - \cos \theta)$
   - (iii) $r = 6/(1 - \cos \theta)$
   - (iv) $r = 2a/(1 + \cos \theta)$
   - (v) $r = 2 + 3 \cos \theta$
   - (vi) $\theta = -\pi/6$
   - (vii) $r = a \cos 2\theta$
   - (viii) $r^2 = a^2 \cos 2\theta$
   - (ix) $r = 2 \sin 3\theta$
   - (x) $r = 1 + 2 \cos \theta$
   - (xi) $r = a \cos^2(\tfrac{1}{2}\theta)$
   - (xii) $r = 1 + \sin \theta$

4. Find the area bounded by $r = a(1 - \cos \theta)$.
5. Find the area bounded by $r = 2(1 + \sin \theta)$.
6. Find the area of the minor segment cut off by the line $x = 1$ from (i) the circle $r = 2$, (ii) the cardioid $r = \tfrac{4}{3}(1 + \cos \theta)$.
7. Find the area of one loop of (i) $r = a \cos 2\theta$, (ii) $r^2 = a^2 \cos 2\theta$.

8. Find the lengths of the following curves:
   (i) $r = a(1 + \cos\theta)$
   (ii) $r = 2a\cos\theta$
   (iii) $r = a$
   (iv) $r = \cos^2(\tfrac{1}{2}\theta)$
9. Show that the curve $r = e^{(\cot\alpha)\theta}$ makes a constant angle $\phi = \alpha$ with the radius vector OP.
10. Show that the curves $r = \sin 2\theta$, $r = \cos\theta$ intersect at the point $\theta = \pi/6$, $r = \tfrac{1}{2}\sqrt{3}$. Find the angle at which they intersect.
11. Show that for the curve $r = a(1 - \cos\theta)$ the angle $\phi = \tfrac{1}{2}\theta$.

Problems 10B

1. Show that $\psi = \theta + \phi$ in figure 10.12. Hence show that
$$\tan\psi = \frac{r\cos\theta + r'\sin\theta}{-r\sin\theta + r'\cos\theta}, \quad \text{where } r' = \frac{dr}{d\theta}$$
   Find ψ at $\theta = \pi/6$ on the curve $r = 2 + \sin\theta$.
2. Find the area of each loop of $r = 1 + 2\cos\theta$.
3. Find the area inside the cardioid $r = a(1 + \cos\theta)$ and outside the circle $r = a$.
4. Find the area inside $r = 2\cos\theta$ and outside $r = 1$.
5. Sketch the curve $r = 1 + \cos 4\theta$ and find the area enclosed by one loop.
6. Sketch the curve $r = a\theta$. If O is the origin and P the point on the curve where $\theta = \tfrac{3}{4}$, find the area bounded by the curve and the radial line $\theta = \tfrac{3}{4}$. Find the length of the arc OP.
7. *Area of revolution.* Show that the area of the surface generated by revolving an arc δs about the x-axis is approximately
$$\delta s . 2\pi y = 2\pi r \sin\theta\, \delta s$$
   Hence show that the area of the surface generated by revolving the arc of $r = f(\theta)$ from $\theta = \alpha$ to $\theta = \beta$ about the x-axis is
$$2\pi \int_\alpha^\beta r \sin\theta\, ds = 2\pi \int_\alpha^\beta r\sin\theta \left\{r^2 + \left(\frac{dr}{d\theta}\right)^2\right\}^{\frac{1}{2}} d\theta$$
   Show that the area of the surface formed by rotating $r = 2a\cos\theta$ about Ox is $4\pi a^2$.

11
Centre of Mass and Centroids

There are some properties of physical bodies which are not directly affected by their size and shape. For instance a body which weighs one kilogram will affect a spring balance the same as any other body with the same weight but of different shape or density. To study these properties it is convenient to consider the total mass of the body as if it were concentrated at a single point, the *centre of mass* of the body. Thus for instance, we may consider the moment of the weight of the body as the moment of the total weight acting through the centre of mass.

Illustrative Example 1

Find the centre of mass of a rod OA of length L, if the mass per unit length at a distance x from one end is $m = kx^2$.

Choosing axes as in figure 11.1, the moment of the weight of a small element δx about O is

$$\text{Force} \times \text{distance} = (m\delta x g)x$$

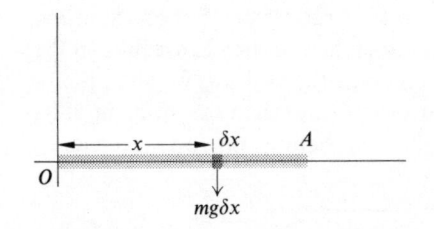

FIGURE 11.1 *Total moment about* $0 = \int_0^L xmg\,dx$

Hence the total moment of the rod about O is

$$\int_0^L mgx\,dx = \int_0^L kgx^3\,dx = \tfrac{1}{4}kgL^4$$

Now, we also know that the moment of the rod about O is the moment of the total weight acting through the centre of mass G, since this is the defining property of G. Therefore $\tfrac{1}{4}kgL^4 = Mgx_G$ where M is the mass of the rod. But

$$M = \int_0^L m\,dx = \int_0^L kx^2\,dx = \tfrac{1}{3}kL^3$$

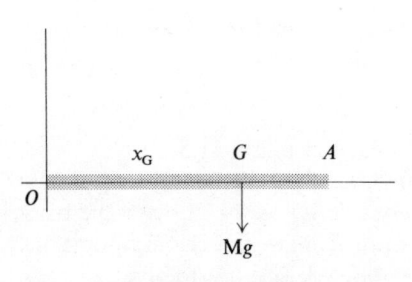

FIGURE 11.2 *Total moment about* $0 = Mgx_G$

Therefore

$$\tfrac{1}{4}kgL^4 = \tfrac{1}{3}kL^3\,gx_G \quad \text{and} \quad x_G = \tfrac{3}{4}L$$

When we are not concerned with the density of any particular body but only with its shape, e.g. a rod, a semicircle, a sphere; we refer to its geometric 'centre' as the *centroid* of the body, i.e. the centroid of a body is the same as the centre of mass when the body has uniform density.

Illustrative Example 2

Find the centroid of a rod of length L.

Taking axes as in example 1 and assuming the mass per unit length m is constant, so that the centroid coincides with the centre of mass, we have by taking moments:

$$\int_0^L mgx\, dx = Mg\bar{x}$$

where $\bar{x}$ is the coordinate of the centroid and

$$M = \int_0^L m\, dx = mL$$

Therefore

$$\tfrac{1}{2}mgL^2 = mLg\bar{x} \quad \text{and} \quad \bar{x} = \tfrac{1}{2}L$$

This result is as expected for the position of the geometric centre of a rod, but the method clearly illustrates the fact that since m is constant the factor mg cancels out at the end of the working. For this reason it is customary to leave out all reference to the weight mg from the beginning of the calculation when finding centroids. In the two-dimensional problems on centroids that follow, we shall therefore talk about the moment of an area rather than the moment of its weight. Hence, we make the following definitions for moments of areas:

$$\boxed{\begin{array}{l}\bar{x}A = x\text{-moment of the area } A \\ \bar{y}A = y\text{-moment of the area } A\end{array}}$$

where $\bar{x}$ and $\bar{y}$ are the coordinates of the centroid of A.

Illustrative Example 3

Find the centroid of the shaded area A in figure 11.3.

First we find the x- and y-moments of the elemental strip shown in figure 11.3. Since it may be considered as a rod, we have from example 1 that its centroid C is approximately at the midpoint, i.e. at $(x, \tfrac{1}{2}y)$. Hence we may use the above definitions to give:

x-moment of strip $\approx x(\text{area of strip}) = xy\,\delta x$
y-moment of strip $\approx \tfrac{1}{2}y(\text{area of strip}) = \tfrac{1}{2}y^2\,\delta x$

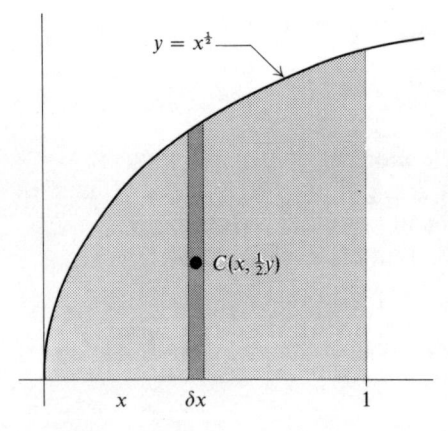

FIGURE 11.3

Now we may sum over a number of such strips and let $\delta x \to 0$, to obtain the x- and y-moments of A:

$$x\text{-moment of } A = \int_0^1 xy\, dx = \int_0^1 x^{3/2}\, dx = [\tfrac{2}{5}x^{5/2}]_0^1 = \tfrac{2}{5}$$

$$y\text{-moment of } A = \int_0^1 \tfrac{1}{2}y^2\, dx = \int_0^1 \tfrac{1}{2}x\, dx = [\tfrac{1}{4}x^2]_0^1 = \tfrac{1}{4}$$

We use the definitions again to deduce the coordinates $(\bar{x}, \bar{y})$ of the centroid of A:

$$\bar{x}A = x\text{-moment of } A = \tfrac{2}{5}$$
$$\bar{y}A = y\text{-moment of } A = \tfrac{1}{4}$$

Since

$$A = \int_0^1 x^{1/2}\, dx = [\tfrac{2}{3}x^{3/2}]_0^1 = \tfrac{2}{3}$$

this gives:

$$\bar{x} = \tfrac{2}{5} / \tfrac{2}{3} = \tfrac{3}{5}, \qquad \bar{y} = \tfrac{1}{4} / \tfrac{2}{3} = \tfrac{3}{8}$$

Class Discussion Exercises 11

1. Find the centroid of the semi-circular area inside the circle $x^2 + y^2 - 2ax = 0$ and above the x-axis.
2. Find the centroid of the volume of revolution formed when the area between $y^2 = x$ and $x = 1$ is rotated through π radians about the x-axis. What is the centre of mass of this volume if the mass per unit volume $= kx$?
3B. *Pappus' first theorem.* If an arc of length S is revolved about an axis l which it does not cross, the area of the surface generated is equal to S multiplied by the distance travelled by the centroid of the arc.
    Find the centroid of the semicircular arc $y = \sqrt{(a^2 - x^2)}$ (i) directly, (ii) using Pappus' first theorem.
4. *Pappus' second theorem.* If an area A is revolved about an axis l, which does not cut the area, the volume generated is equal to A multiplied by the distance travelled by the centroid of the area.
    Find the volume of revolution formed when the semicircular area in exercise 1 is rotated through 2π radians about the y-axis.

Problems 11A

1. Find the centroid of the triangular area between $y = mx$, $x = h$, $x = 0$. Deduce by using Pappus' theorem the volume of a right circular cone of vertical height h and base radius r.

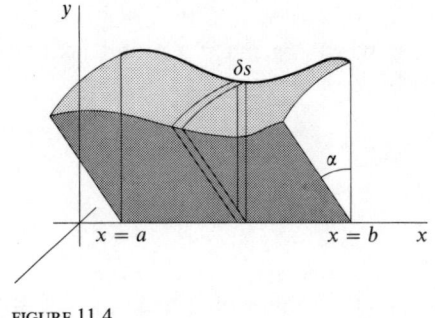

FIGURE 11.4

2. Find the volume of the torus obtained by rotation of the circle $x^2 + y^2 = 4$ about the line $x = 4$.

3. A rectangle of sides a and h is rotated about a side of length h. Show by Pappus' theorem that the total surface area of the cylinder generated is $\pi a(2a + h)$, and that its volume is $\pi d^2 h$.

4. Find the centroids of the areas bounded by the following curves and find the volumes of revolution about the lines indicated.
   (i) $y = x^{\frac{3}{2}}$, $x = 1$, $y = 0$; Vol. about Ox
   (ii) $y^2 = 3x$, $x = 3$; Vol. about $x = -1$
   (iii) $y = x^2$, $y = 1$; V_{Ox} (Express as an integral in y)
   (iv) $x = e^y$, $y = 0$, $y = 1$, $x = 0$; V_{Ox}, V_{Oy}
   (v) $2y = x$, $2y = x^2$
   (vi) $y^3 = x$, $y = x^2$

5. Use Pappus' theorem to find the position of the centroid of the first quadrant area inside the circle $x^2 + y^2 = a^2$. Find the volume of revolution of this area about the line $x + y = 1$.

6. Find the centroid of the solids formed when the plane areas bounded by the following curves are rotated about the given axes:
   (i) $y^2 = x$, $x = 3$, $y = 0$; x-axis
   (ii) $y^2 = x$, $x = 3$, $y = 0$; y-axis
   (iii) $y = x^{\frac{3}{2}}$, $x = 1$, $y = 0$; x-axis
   (iv) $y = mx$, $x = 0$, $x = h$, $y = 0$; x-axis
   (v) $x = e^y$, $y = 0$, $y = 1$, $x = 0$; y-axis

7. Find the centre of mass of a rod of length a and mass per unit length proportional to $x(a - x)$, where x is the distance from one end.

8. Find the centre of mass of the lamina formed by the area between the curves $y = x$ and $x = 1$, when the mass per unit area is proportional to the distance from the y-axis.

Problems 11B

1. Sketch the area bounded by the curves $y^2 = 6x$ and $x^2 = 6y$. Find the centroid of the area and the volume obtained by rotating it about the x-axis.

2. The region $0 \leq x \leq \frac{1}{2}\pi$, $0 \leq y \leq \cos x$ is rotated (i) about the x-axis, (ii) about the y-axis. Find by integration the volumes of the solids of revolution generated, and use a theorem of Pappus to deduce the coordinates of the centroid of the region.

3. Find the centre of mass of the semi-circular plate $x^2 + y^2 < a^2$, $y > 0$, if the mass per unit area is proportional to y.

4. Find the x coordinate of the centroid of the arc of $y = \frac{2}{3}x^{\frac{3}{2}}$ between $x = 0$ and $x = 3$. Find the area of the surface generated when this arc is rotated about the y-axis.

5. Find the centroid of the arch of the curve $x = a(\theta - \sin\theta)$, $y = a(1 - \cos\theta)$ from $\theta = 0$ to 2π.
6. Find the centroid of the first quadrant arc of $x = a\cos^3\theta$, $y = a\sin^3\theta$.
7. A sector of a circle with radius a has angle 2α. Find the position of its centroid of area.

12
Moments of Inertia

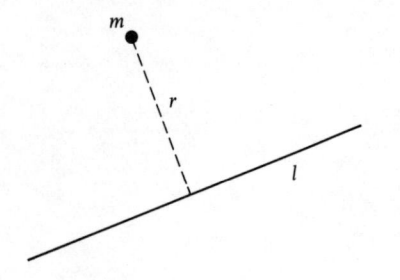

FIGURE 12.1

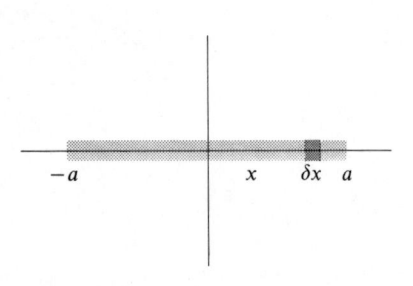

FIGURE 12.2

The *moment of inertia* of a particle of mass m about an axis l is defined as:

$$I_l = mr^2$$

where r is the perpendicular distance from the particle to the axis. For a collection of particles m_i ($i = 1, 2, \ldots, n$) distant r_i from the axis l, we have:

$$I_l = \sum_{i=1}^{n} m_i r_i^2$$

Illustrative Example 1

Find the moment of inertia of a thin rod of length $2a$ about an axis l through the centre of the rod and perpendicular to it, if the mass per unit length is a constant, m.

First we choose axes as in figure 12.2, so that the rod extends from $x = -a$ to $x = a$. The axis l is now the y-axis. Considering the element shown as a particle of mass $m\,\delta x$, we have for the moment of inertia of the element about Oy:

$$I_{Oy} \approx (m\,\delta x)x^2$$

Summing over all elements of the rod and letting $\delta x \to 0$, this gives for the rod:

$$I_{Oy} = \int_{-a}^{a} mx^2\,dx = \tfrac{2}{3}ma^3$$

It is often useful to write the moment of inertia in the form Mk^2, where M is the mass and k is called the *radius of gyration*. Hence we have for the rod, where $M = 2ma$:

$$I_l = \tfrac{1}{3}Ma^2$$

and the radius of gyration is $a/\sqrt{3}$.

Moments of Inertia 71

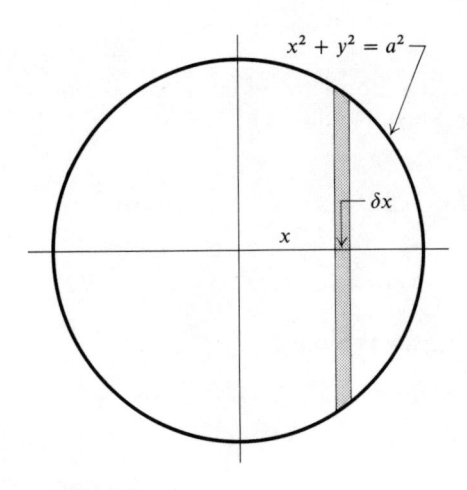

FIGURE 12.3

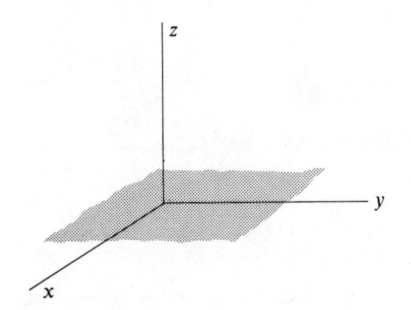

FIGURE 12.4

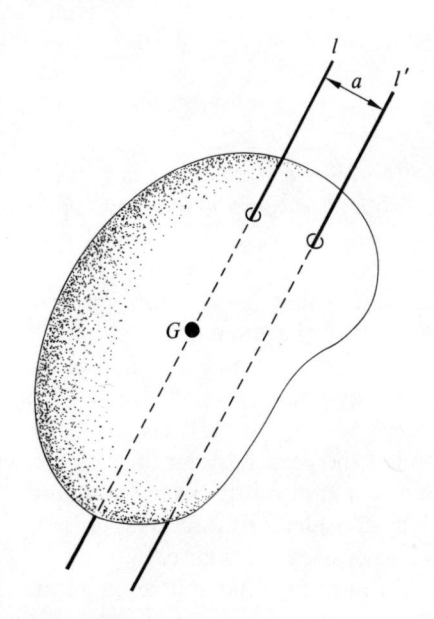

FIGURE 12.5

Illustrative Example 2

Find I_l for a disc of uniform mass per unit area m and radius a, where l is a diameter of the disc.

Choose axes as in figure 12.3 so that l becomes the x-axis. Considering the elemental strip as a rod of approximate length $2y$ and mass $y\delta x\, m$, we may use the result of example 1 to give:

$$I_{Ox} \approx \tfrac{1}{3}(2y\delta x\, m)y^2$$

Therefore, summing over the whole disc and letting $\delta x \to 0$, we have for the whole disc:

$$I_{Ox} = \tfrac{2}{3}m \int_{-a}^{a} y^3\, dx = \tfrac{4}{3}m \int_{0}^{a} (a^2 - x^2)^{\tfrac{3}{2}}\, dx$$

Putting $x = a \sin \theta$:

$$I_{Ox} = \tfrac{4}{3}ma^4 \int_{0}^{\pi/2} \cos^4 \theta\, d\theta$$

and by Wallis' formula (see Chapter 9):

$$I_{Ox} = \frac{4}{3}ma^4 \cdot \frac{3}{4} \cdot \frac{1}{2} \cdot \frac{\pi}{2} = \tfrac{1}{4}\pi ma^4$$

Since $M = \pi a^2 m$, this gives:

$$I_l = \tfrac{1}{4}Ma^2$$

and the radius of gyration: $k = \tfrac{1}{2}a$.

Perpendicular Axes Theorem For bodies in the form of *plane areas*, we prove in class discussion exercise 2 that if we choose axes so that Ox and Oy are in the plane of the body (figure 12.4), then

$$I_{Oz} = I_{Ox} + I_{Oy}$$

Parallel Axes Theorem We prove in class discussion exercise 3 that if I_l is the moment of inertia of a body about an axis l which passes through its centre of mass and l' is an axis parallel to l then

$$I_{l'} = I_l + Ma^2$$

where M is the mass of the body and a the perpendicular distance between l and l'.

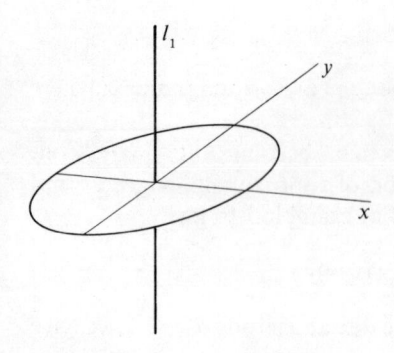

FIGURE 12.6(a)

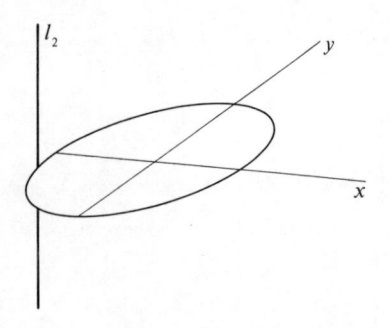

FIGURE 12.6(b)

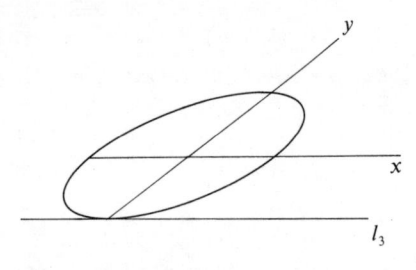

FIGURE 12.6(c)

Illustrative Example 3

Find the moment of inertia of a uniform disc of mass M and radius a about each of the three axes shown in figure 12.6.

(i) From example 2 we have
$$I_{Ox} = \tfrac{1}{4}Ma^2$$
and by symmetry
$$I_{Oy} = \tfrac{1}{4}Ma^2$$
Therefore by the perpendicular axes theorem
$$I_{l_1} = I_{Oz} = I_{Ox} + I_{Oy}$$
$$= \tfrac{1}{4}Ma^2 + \tfrac{1}{4}Ma^2 = \tfrac{1}{2}Ma^2$$

(ii) By the parallel axes theorem
$$I_{l_2} = I_{l_1} + Ma^2$$
Therefore
$$I_{l_2} = \tfrac{1}{2}Ma^2 + Ma^2 = \tfrac{3}{2}Ma^2$$

(iii) Again by the parallel axes theorem we have:
$$I_{l_3} = I_{Ox} + Ma^2$$
$$= \tfrac{1}{4}Ma^2 + Ma^2 = \tfrac{5}{4}Ma^2$$

Class Discussion Exercises 12

1. Find the moment of inertia of a uniform rectangular lamina of sides $2a$ and $2b$ about an axis through the centroid and parallel to the edge of length $2a$.
2. Prove the perpendicular axes theorem. Find the moment of inertia of the lamina in exercise 1 about an axis through the centroid perpendicular to the plane of the lamina.
3. Prove the parallel axes theorem. Find the moment of inertia of the lamina in exercise 1 about the axes:
   (i) along the edge of length $2a$,
   (ii) through a corner of the lamina and perpendicular to its plane.
4. Find the moment of inertia of a thin uniform ring of mass M and radius a about its axis. Hence by considering a disc as a series of concentric rings find the moment of inertia of a disc.
5. Find the moment of inertia of a uniform solid sphere, about a diameter.

REFERENCE TABLE FOR MOMENTS OF INERTIA OF UNIFORM BODIES

Body	Dimension	Position of axis	I
Rod	length $2a$	through one end, perpendicular to rod	$\frac{4}{3}Ma^2$
		through centre, perpendicular to rod	$\frac{1}{3}Ma^2$
Ring	radius a	along diameter	$\frac{1}{2}Ma^2$
		through centre perpendicular to plane of ring	Ma^2
Hollow cylinder	radius a	along main axis	Ma^2
Disc	radius a	along diameter	$\frac{1}{4}Ma^2$
		through centre perpendicular to plane of disc	$\frac{1}{2}Ma^2$
Solid cylinder	radius a	along main axis	$\frac{1}{2}Ma^2$
Solid sphere	radius a	along diameter	$\frac{2}{5}Ma^2$
Hollow sphere	radius a	along diameter	$\frac{2}{3}Ma^2$

Problems 12A

1. Show that the moment of inertia of a solid circular cylinder of mass M, radius a and length l about a diameter of an end face is $M(a^2/4 + l^2/3)$.
2. Find by integration the moment of inertia of a hollow cylinder of length l and radius a about its axis.
3. A metal drum closed at both ends of radius one metre and length two metres is made of uniform gauge steel. Find the radius of gyration about its axis.
4. Find the moment of inertia of a circular ring of mass M and radius a
   (i) about an axis through a point of the ring perpendicular to its plane,
   (ii) about a diameter.
5. Find the moment of inertia of a square lamina of side $2a$ about a diagonal.
6. Find the moment of inertia of a square wire frame of side $2a$
   (i) about an axis along one side of the frame,
   (ii) about an axis through one corner, perpendicular to the plane of the frame.
7. Find the moment of inertia of a triangular lamina of height h and base a about an axis parallel to the base through the vertex.
8. Find the radius of gyration of the I piece shown in figure 12.7:
   (i) about the axis XX', (ii) about the axis YY'. (All the measurements are in centimetres.)

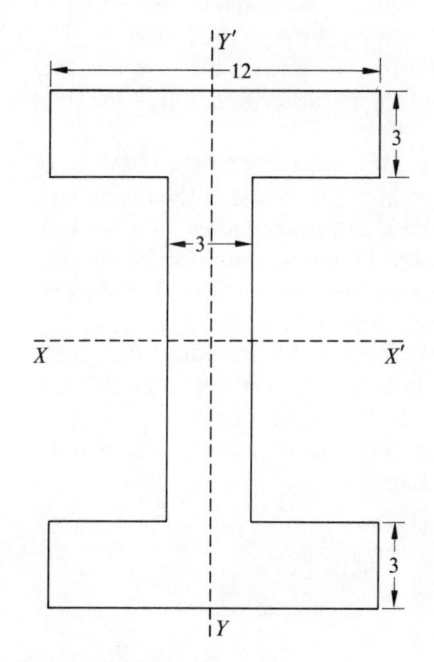

FIGURE 12.7

74 *Moments of Inertia*

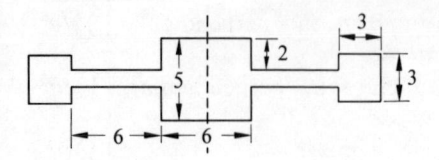

FIGURE 12.8

9. Find the moment of inertia of a thin spherical shell of mass M and radius a. Hence by considering a sphere as a series of concentric shells find the moment of inertia of a uniform solid sphere of mass M and radius a about a diameter.
10. Show that the moment of inertia of a uniform solid ring, of rectangular cross-section, height h and internal and external radii of a and b respectively about its axis is $\frac{1}{2}M(a^2 + b^2)$. Find the radius of gyration of the flywheel, whose cross-section is given in figure 12.8, about its central axis. (All dimensions are in centimetres.)

Problems 12B

1. Show that the radius of gyration of a uniform solid circular cone of height h and radius of base r, about a diameter of the base is given by $k^2 = \frac{1}{10}h^2 + \frac{3}{20}r^2$.
2. Show that the moment of inertia of a thin uniform rod of length $2a$ and mass M about an axis through its centre inclined at an angle θ to the rod is $\frac{1}{3}Ma^2 \sin \theta$. Deduce that the moment of inertia of the rod about any line in space is the same as for a weightless rod of length $2a$ with masses $\frac{1}{6}M$ attached at each end and a mass $\frac{2}{3}M$ attached at the centre.
3. A lamina of uniform density ρ is formed by one loop of the curve $r = 2 \sin 2\theta$. Find the moment of inertia of the lamina about an axis normal to its plane and passing through the origin.
4. The area between the parabola $y = x^2$ and the line $y = x$ is rotated about the y-axis. Find the radius of gyration of the solid formed, about the y-axis.
5. An astronomer uses the value $\frac{1}{3}Ma^2$ for the moment of inertia of the Earth about the axis of rotation, where M is the mass and a the radius. Assuming that the Earth is composed of a uniform outer shell of depth $\frac{1}{2}a$ and mass per unit volume m and a uniform central core of radius $\frac{1}{2}a$ and mass per unit volume m', show that the ratio $m':m = 23:7$, if the astronomer's value is correct.
6. A toroid (anchor ring) is formed by rotating the circle $(x - l)^2 + y^2 = r^2$, $l > r$, about the y-axis through 2π radians. Show that the moment of inertia about the y-axis is $M(l^2 + \frac{3}{4}r^2)$.
7. Show that the moment of inertia of a rectangular lamina of mass M and sides $2a$, $2b$ about a diagonal is

$$\frac{2Ma^2b^2}{3(a^2 + b^2)}$$

13

Numerical Integration

In many cases where integrals arise, they occur not as indefinite integrals which must be evaluated by the formal techniques of integration, but as definite integrals where the limits of integration are known; so that we only need to find a numerical answer to some desired degree of accuracy. As we know, the definite integral $\int_a^b f(x)\,dx$ may be regarded as the area beneath $y = f(x)$ between $x = a$ and $x = b$, so that the problem of calculating the numerical value of the integral is equivalent to that of finding an area. This is the approach we adopt in deriving two rules for numerical evaluation of definite integrals. The first of these is the *trapezoidal rule*, which is to some extent of academic interest only since the other rule derived, *Simpson's rule*, is much more accurate and is a standard program in every computer library.

Illustrative Example 1

Find an approximate value for $\int_1^7 \sqrt{(3+x)}\,dx$.

Here we have to estimate the area shown under $y = \sqrt{(3+x)}$ between $x = 1$ and $x = 7$. The formula which we use for this estimate is derived by dividing the area into n strips of width h. In this example we shall take $n = 6$, so that the interval between $x = 1$ and $x = 7$ is divided into six strips of width $h = 1$. Now we label points on the x-axis separated by the length $h = 1$ so that:

$$x_0 = 1, \quad x_1 = 2, \quad x_2 = 3, \quad x_3 = 4, \quad x_4 = 5, \quad x_5 = 6, \quad x_6 = 7$$

The corresponding y-coordinates may now be calculated from the formula $y = \sqrt{(3+x)}$, so that $y_0 = \sqrt{(3+1)}$, $y_1 = \sqrt{(3+2)}$, etc.:

y_0	y_1	y_2	y_3	y_4	y_5	y_6
2	2·2361	2·4495	2·6458	2·8284	3·0000	3·162

We now approximate the area of the first strip by finding the area of the trapezium beneath the *chord* P_0P_1, rather than the arc P_0P_1. The

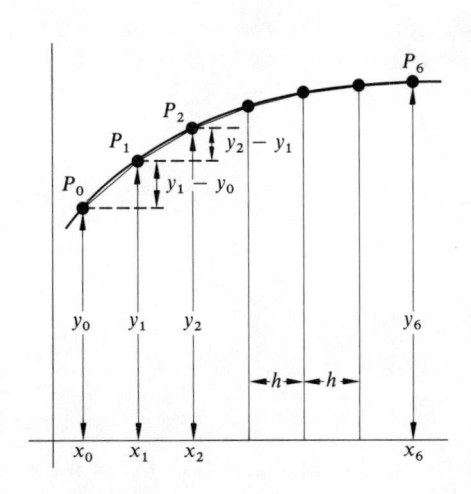

FIGURE 13.1

Numerical Integration

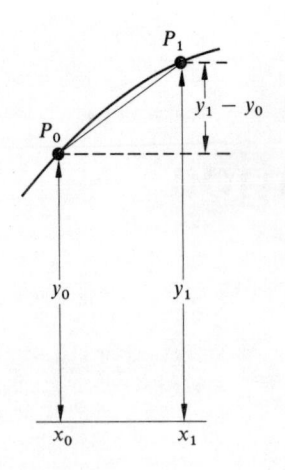

FIGURE 13.2

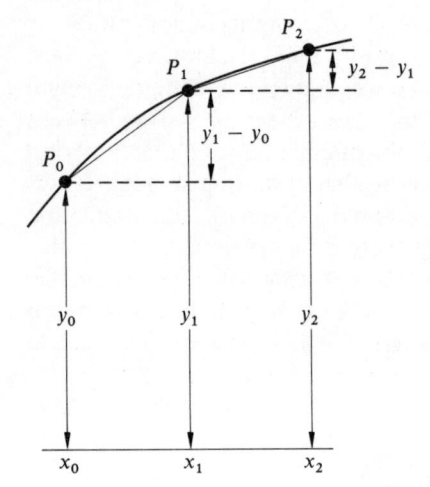

FIGURE 13.3

area of this trapezium may be considered as the sum of a rectangle plus a small triangle of height $(y_1 - y_0)$, i.e.:

$$\text{Area of trapezium} = y_0 h + \tfrac{1}{2}(y_1 - y_0)h = \tfrac{1}{2}(y_0 + y_1)h$$

Treating the other trapezia in the same way, we have:

$$\int_1^7 \sqrt{(3 + x)}\, dx \approx \tfrac{1}{2}h(y_0 + y_1) + \tfrac{1}{2}h(y_1 + y_2) + \cdots + \tfrac{1}{2}h(y_5 + y_6)$$
$$= \tfrac{1}{2}h(y_0 + 2y_1 + 2y_2 + \cdots + 2y_5 + y_6)$$
$$= \tfrac{1}{2}\cdot 1(2 + 2(13\cdot 147) + 3\cdot 162) = 15\cdot 728$$

The above formula is known as the trapezoidal rule, and is given below for the general case of n strips of width h:

TRAPEZOIDAL RULE

If $y = f(x)$,

$$\int_a^b f(x)\, dx \approx \tfrac{1}{2}h\{y_0 + 2(y_1 + y_2 + \cdots + y_{n-1}) + y_n\}$$

where the interval $a \leq x \leq b$ is divided into n equal parts of width h.

By dividing the area under $y = f(x)$ into an *even* number of strips, the much more accurate Simpson's rule may be obtained. Considering two strips at a time, we may derive an approximation to the area by using the first three terms in the Taylor expansion of $f(x)$. (The trapezoidal rule is equivalent to taking the first *two* terms.)

We have for the first two strips (figure 13.3), a total area A given by

$$A = \int_{x_0}^{x_2} f(x)\, dx$$

Putting $x = x_1 + t$, we may convert this integral to one in the variable t, where the limits for t are $t = -h$ and $t = +h$.

$$A = \int_{-h}^{h} f(x_1 + t)\, dt = \int_{-h}^{h} \left\{ f(x_1) + \frac{f'(x_1)}{1!} t \right.$$
$$\left. + \frac{f''(x_1)}{2!} t^2 + \cdots \right\} dt$$
$$\approx \left[f(x_1)t + f'(x_1)\frac{t^2}{2} + \frac{f''(x_1)}{2}\frac{t^3}{3} \right]_{-h}^{h}$$

Therefore

$$A \approx 2hf(x_1) + \tfrac{1}{3}h^3 f''(x_1)$$
$$= 2hy_1 + \tfrac{1}{3}h^3 f''(x_1)$$

In this expression we must estimate the value of $f''(x_1)$, i.e. we must estimate the *rate of change of the gradient of the tangent*. Since this is approximately the same as the rate of change of the gradient of the *chord*, we may consider chords P_0P_1 and P_1P_2 and write:

$$f''(x_1) \approx \{(\text{gradient } P_1P_2) - (\text{gradient } P_0P_1)\}/h$$
$$= \left\{\frac{(y_2 - y_1)}{h} - \frac{(y_1 - y_0)}{h}\right\}\bigg/h$$
$$= (y_2 - 2y_1 + y_0)/h^2$$

Hence we have:

$$A \approx 2hy_1 + \tfrac{1}{3}h(y_2 - 2y_1 + y_0) = \tfrac{1}{3}h(y_2 + 4y_1 + y_0)$$

Since similar formulae apply for the areas of subsequent pairs of strips, we obtain:

$$\int_a^b f(x)\,dx \approx \tfrac{1}{3}h(y_0 + 4y_1 + y_2) + \tfrac{1}{3}h(y_2 + 4y_3 + y_4)$$
$$+ \cdots + \tfrac{1}{3}h(y_{2n-2} + 4y_{2n-1} + y_{2n})$$

Collecting terms together we obtain:

SIMPSON'S RULE

$$\int_a^b f(x)\,dx = \tfrac{1}{3}h[y_0 + 4(y_1 + y_3 + \cdots) + 2(y_2 + y_4 + \cdots) + y_{2n}]$$

$$= \tfrac{1}{3}h[(\text{first + last ordinate}) + 4 \text{ times (sum of odd ordinates)}$$
$$+ 2 \text{ times (sum of the remaining even ordinates)}]$$

Illustrative Example 2

Use Simpson's rule to evaluate the definite integral: $\int_1^7 \sqrt{(3 + x)}\,dx$ with 6 strips of width $h = 1$.

The calculation may be tabulated as follows:

x	$\sqrt{(3 + x)}$	first and last y_0, y_6	odd $y_1, y_3, \ldots$	even $y_2, y_4, \ldots$
$x_0 = 1$	2	2		
$x_1 = 2$	2·2361		2·2361	
$x_2 = 3$	2·4495			2·4495
$x_3 = 4$	2·6458		2·6458	
$x_4 = 5$	2·8284			2·8284
$x_5 = 6$	3		3	
$x_6 = 7$	3·1623	3·1623		
		5·1623	7·8819	5·2779
			×4	×2
		5·1623 +	31·5276 +	10·5558 = 47·2457

Therefore
$$\int_1^7 \sqrt{(3 + x)}\, dx \approx (47\cdot 2457)(\tfrac{1}{3}) = 15\cdot 749$$

Class Discussion Exercises 13

1. Evaluate approximately:
$$\int_0^{0\cdot 4} \sqrt{(1 - x^2)}\, dx$$
using (i) the trapezoidal rule, (ii) Simpson's rule, with 4 strips. Check the accuracy by direct integration.

2. Evaluate the integral in exercise 1 approximately, by expanding the integrand as a Maclaurin series. Why is this method not applicable to the integral in illustrative example 1?

3. Evaluate by direct integration
$$\int_1^7 \sqrt{(3 + x)}\, dx$$
Compare the accuracies of the results in illustrative examples 1 and 2.

4. A smooth curve is drawn through points with x and y coordinates:

x	2	3	4	5	6	7	8
y	6	8·3	10·1	10·1	11·2	12·5	13·5

Find the approximate area between this section of the curve and the x-axis.

5B. An estimate of the error E involved in using Simpson's rule for $\int_a^b f(x)\, dx$ may be made by using the formula:
$$|E| < \frac{h^4(b - a)}{180} \cdot |\max f^4(x), \ a \leqslant x \leqslant b|$$

If the calculations in illustrative example 2 are performed to seven decimal places accurately, the Simpson's rule estimate for $\int_1^7 \sqrt{(3 + x)}\, dx$ using 4 strips gives 15·748 462 8. The correct answer is 15·748 517 7 to seven decimal places. Use these figures to verify the above formula.

By considering the formula, say by what factor you expect the error $|E|$ to decrease when the number of strips in Simpson's rule is doubled. The Simpson's rule estimate for the above integral using 12 strips is 15·748 514 1. By what factor has $|E|$ decreased?

Problems 13A

1. Evaluate the following integrals to three significant figures, using Simpson's rule with 4 strips in each case:

   (i) $\int_2^4 \frac{1}{x} dx$ (ii) $\int_0^{\pi/2} \sin x \, dx$ (iii) $\int_1^2 \log_e x \, dx$

   (iv) $\int_0^{\pi/4} \tan x \, dx$ (v) $\int_0^1 \sqrt{(1+2x^2)} \, dx$ (vi) $\int_0^1 (x^2+4)^{-\frac{1}{2}} \, dx$

2. Evaluate $\int_0^1 \sqrt{(4+x^4)} \, dx$ to three significant figures by Simpson's rule with 4 strips. Check by expanding the integrand up to the term in x^8.

3. Evaluate
$$\int_2^3 \frac{dx}{\log_{10} x}$$
by Simpson's rule using 4 strips.

4. Evaluate
$$\int_0^4 \frac{dx}{\sqrt{(64-x^2)}}$$
to four decimal places using Simpson's rule with 8 strips. Obtain the integral by direct integration and hence estimate π to three decimal places.

5. By expanding the integrand and separately by using Simpson's rule with 4 strips, evaluate correct to three decimal places.
$$\int_0^{\frac{1}{2}} \frac{dx}{\sqrt{(1-x^3)}}$$

6. Find the approximate area between the x-axis, $x = 0$, $x = 6$, and a smooth curve through the points with x- and y-coordinates:

x	0	1	2	3	4	5	6
y	8	8·7	9	8	6·3	5·2	4·7

7. The area under the curve $y = \log_e x$ from $x = 1$ to $x = 2$ is revolved about the x-axis. Find an approximation to the volume of the resulting solid using Simpson's rule with 4 strips.

Problems 13B

1. Show that Simpson's rule with two strips gives the exact answer to $\int_0^2 x^3 \, dx$. Explain why the answer is exact, with reference to the derivation of Simpson's rule given in illustrative example 2.

2. The speed of a car v metres per second at three-second intervals after starting from rest at time $t = 0$ is recorded from the speedometer as:

t	0	3	6	9	12	15	18
v	0	3	8	13	19	24	28

Estimate the distance travelled in 18 seconds.

3. An object is dropped in an atmosphere producing a resistance proportional to $v^{\frac{3}{2}}$, where v is its velocity. The time taken for it to reach a velocity v is given by:

$$t = \frac{u}{g} \int_0^{v/u} \frac{1}{(1 - x^{\frac{3}{2}})} dx$$

where u is the terminal velocity. Find to three significant figures the time taken for it to reach a velocity of $\frac{1}{2}u$.

4. Use Simpson's rule with two strips to evaluate the integral $\int_0^1 (2x^2 - 1) \, dx$. Compare your answer with the exact result.

5. Show from the formula in class discussion exercise 5 that the error in using Simpson's rule in problem A1(i) is less than $h^4/120$. Hence show that the error E in the answer to A1(i) is numerically less than 0·001. How many strips should we take if we need the error to be less than 0·000 000 1?

6. In evaluating a certain definite integral by Simpson's rule using 4 strips, the result 2·133 142 56 was obtained. The integral was then evaluated by taking 40 strips, and the result 2·132 492 26 obtained. To how many places of decimals was the first result accurate? To how many decimal places is the second result likely to be reliable?

14
Fourier Series

A very great number of the functions arising in physical problems are of the type known as *periodic functions*. That is to say: their graphs repeat their form over a certain interval in the next and subsequent intervals, as in figures 14.1 and 14.2. The simplest periodic

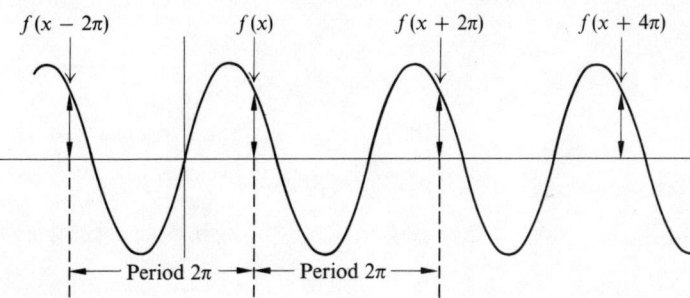

FIGURE 14.1

functions are those of sine and cosine: if we look at figure 14.1, we see that the $y = \sin x$ graph repeats itself every time that x increases by 2π, and so we say that $\sin x$ is periodic, with period 2π.

In general, $f(x)$ is periodic with period T if for all x, $f(x + T) = f(x)$. Other functions, with different periods are illustrated in figure 14.2.

In the study of Fourier series, we are concerned with the trigonometric functions:

$$\sin x, \sin 2x, \sin 3x, \ldots, \sin nx; \quad \cos x, \cos 2x, \ldots, \cos nx$$

and with the sort of functions that we get by adding them together. Notice that whenever any of these functions, multiplied by constants, are added together, the sum always has the period 2π. For instance,

$$f(x) = \sin x - \tfrac{1}{2} \sin 2x + \tfrac{1}{3} \sin 3x - \tfrac{1}{4} \sin 4x + \cdots$$

has period 2π, since $\sin(x + 2\pi) = \sin x$, $\sin 2(x + 2\pi) = \sin 2x$, etc. Therefore $f(x + 2\pi) = f(x)$.

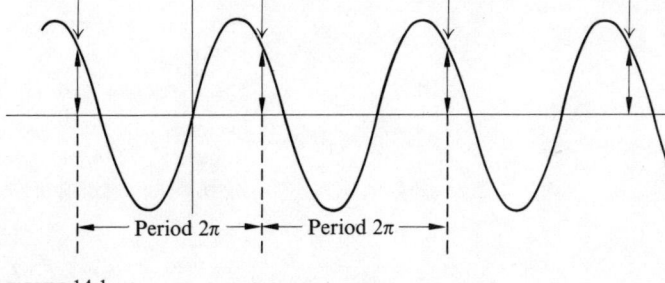

FIGURE 14.2(a)

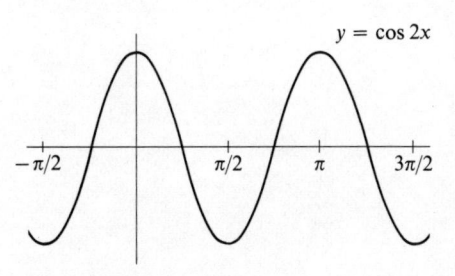

FIGURE 14.2(b)

82 Fourier Series

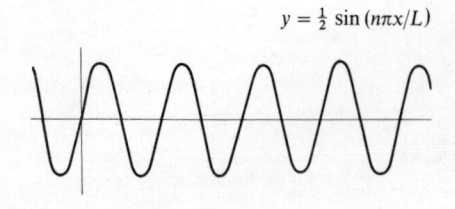

$y = \frac{1}{2}\sin(n\pi x/L)$

FIGURE 14.2(c)

Figure 14.3 shows successive approximations to the graph of $f(x) = \sin x - \frac{1}{2}\sin 2x + \frac{1}{3}\sin 3x - \frac{1}{4}\sin 4x + \cdots$. We may see that the graph of the sum of 20 terms in the series is becoming practically indistinguishable on the scale that we have drawn it from the saw-tooth function in figure 14.3. By taking even more terms in the series, we can get as close as we like to this function. In fact, for the sum to infinity:

$$f(x) = \sum_{n=1}^{\infty} \frac{(-1)^{n+1} \sin nx}{n} = \sin x - \frac{1}{2}\sin 2x + \frac{1}{3}\sin 3x - \cdots \tag{14.1}$$

the graph is *actually* the same as figure 14.3. Therefore the saw-tooth function is represented exactly by the infinite sum (14.1), which is called the *Fourier series* of the function.

It is a surprising fact that *any* function $f(x)$ may be expressed as a Fourier series, as long as:
(1) $f(x)$ is periodic with period $2L$,
(2) There are only a finite number of finite discontinuities and maxima and minima in one period of $f(x)$.

The Fourier series is:

$$f(x) = \tfrac{1}{2}a_0 + \sum_{n=1}^{\infty} \left(a_n \cos \frac{n\pi x}{L} + b_n \sin \frac{n\pi x}{L} \right)$$

where the *Fourier coefficients* a_n and b_n may be found by using the formulae:

$$a_n = \frac{1}{L}\int_{-L}^{L} f(x) \cos \frac{n\pi x}{L} \, dx$$

$$b_n = \frac{1}{L}\int_{-L}^{L} f(x) \sin \frac{n\pi x}{L} \, dx$$

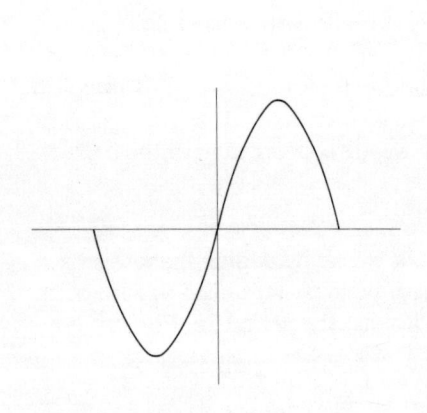

1 term: $\sin x$

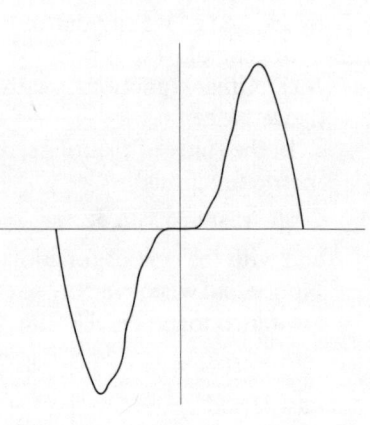

2 terms: $\sin x - \frac{1}{2}\sin 2x$

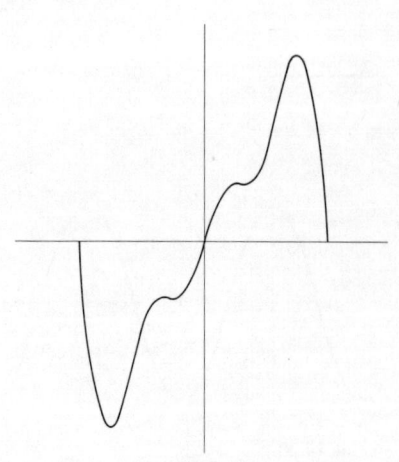

3 terms: $\sin x - \frac{1}{2}\sin 2x + \frac{1}{3}\sin 3x$

FIGURE 14.3 *Fourier series approximations to the saw-tooth function*

Illustrative Example 1

Find the Fourier series for the function defined as periodic with period 2π, such that $f(x) = \tfrac{1}{2}x$ for $-\pi < x < \pi$ (figure 14.4). In this example the period is 2π, hence $2L = 2\pi$, therefore $L = \pi$. For functions with this convenient period the arguments of the sines and cosines involved are less complicated, since for example $\cos(n\pi x)/L$ becomes $\cos nx$, etc. Using the formulae,

$$a_n = \frac{1}{\pi} \int_{-\pi}^{\pi} \tfrac{1}{2}x \cos nx \, dx$$

Integrating by parts,

$$a_n = \frac{1}{\pi}\left[\tfrac{1}{2}x \frac{\sin nx}{n}\right]_{-\pi}^{\pi} - \frac{1}{n\pi}\int_{-\pi}^{\pi} \tfrac{1}{2}\sin nx \, dx$$

$$= \left[x\frac{\sin nx}{2n\pi} + \frac{\cos nx}{2n^2\pi}\right]_{-\pi}^{\pi} = 0$$

The coefficients a_n are zero, but it is usually necessary to check a_0 separately, since the general evaluation of a_n often goes wrong when $n = 0$ (for example: $\int \sin nx \, dx = -(\cos nx)/n$ is not valid if $n = 0$). Hence, separately:

$$a_0 = \frac{1}{\pi}\int_{-\pi}^{\pi} \tfrac{1}{2}x \cos 0 \, dx = \frac{1}{\pi}\int_{-\pi}^{\pi} \tfrac{1}{2}x \, dx = \frac{1}{\pi}\left[\tfrac{1}{4}x^2\right]_{-\pi}^{\pi} = 0$$

and a_0 is also zero.

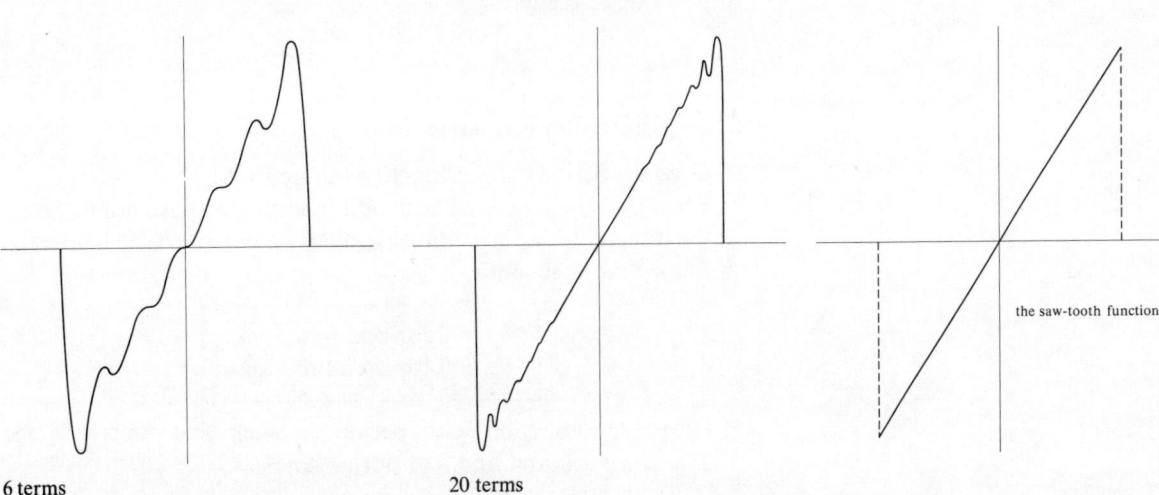

6 terms

20 terms

the saw-tooth function

For the sine coefficients:

$$b_n = \frac{1}{\pi} \int_{-\pi}^{\pi} \tfrac{1}{2} x \sin nx \, dx$$

$$= \frac{1}{2\pi}\left[x \frac{(-\cos nx)}{n} \right]_{-\pi}^{\pi} + \frac{1}{2\pi} \int_{-\pi}^{\pi} \frac{(\cos nx)}{n} dx$$

$$= \frac{1}{2\pi}\left[-x\frac{\cos nx}{n} + \frac{\sin nx}{n^2} \right]_{-\pi}^{\pi}$$

$$= \frac{1}{2n\pi}[-\pi \cos n\pi - \pi \cos(-n\pi)] = -\frac{1}{n}\cos n\pi$$

Noticing that $\cos n\pi = (-1)^n$, this becomes:

$$b_n = \frac{(-1)^{n+1}}{n}$$

Therefore

$$f(x) = \sin x - \tfrac{1}{2}\sin 2x + \tfrac{1}{3}\sin 3x - \tfrac{1}{4}\sin 4x + \cdots$$

Point of Discontinuity In example 1, the function has a discontinuity at $x = \pi$. If we substitute $x = \pi$ in the series, we have:

$$\sin \pi - \tfrac{1}{2}\sin 2\pi + \tfrac{1}{3}\sin 3\pi - \cdots = 0$$

We see that the sum of the Fourier series at a point of discontinuity is the *average* of the values of $f(x)$ on each side. This result is in fact true generally:

At a point of discontinuity $x = a$, the sum of the Fourier series is

$$\boxed{\tfrac{1}{2}\{f(a+0) + f(a-0)\}}$$

where $f(a + 0)$ is the value of $f(x)$ to the right of $x = a$, and $f(a - 0)$ is the value to the left.

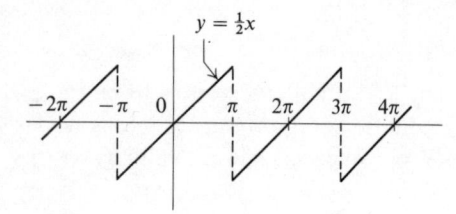

FIGURE 14.4 *Saw-tooth function*

Class Discussion Exercises 14

1. Give the periods of the functions in figure 14.2.
2. Show that since $x \cos nx$ is an odd function, we need not integrate the integral for a_n in example 1, but we can write $a_n = 0$ directly. Show that in general:

$$\boxed{\begin{array}{l} a_n = 0 \text{ for odd functions} \\ b_n = 0 \text{ for even functions} \end{array}}$$

3. Graph the function with period 2π such that $f(x) = x^2$ for $-\pi < x < \pi$, and find its Fourier series.

4. Show that

(i) $\int_{-L}^{L} \sin \dfrac{n\pi x}{L} \sin \dfrac{m\pi x}{L}\, dx = \begin{cases} 0: & m \neq n \\ L: & m = n \neq 0 \end{cases}$

(ii) $\int_{-L}^{L} \cos \dfrac{n\pi x}{L} \cos \dfrac{m\pi x}{L}\, dx = \begin{cases} 0: & m \neq n \\ L: & m = n \neq 0 \end{cases}$

(iii) $\int_{-L}^{L} \sin \dfrac{n\pi x}{L} \cos \dfrac{m\pi x}{L}\, dx = 0$

5. Use the results of exercise 4 to show that if

$$f(x) = \tfrac{1}{2}a_0 + \sum_{n=1}^{\infty} \left\{ a_n \cos\left(\dfrac{n\pi x}{L}\right) + b_n \sin\left(\dfrac{n\pi x}{L}\right) \right\}$$

then a_n and b_n are given by the formulae in the text.

6. Graph the function $f(x) = \begin{cases} 1: & 0 < x < 2 \\ 0: & -2 < x < 0 \end{cases}$, period 4.

Find the Fourier series and show that

$$1 - \tfrac{1}{3} + \tfrac{1}{5} - \tfrac{1}{7} + \cdots = \dfrac{\pi}{4}$$

7. Show that

$$\cos\left(\dfrac{n\pi x}{L}\right) = \cos\left\{\dfrac{n\pi(x + 2L)}{L}\right\}$$

and that if $f(x)$ is of period $2L$, then so is $f(x).\cos(n\pi x/L)$. By considering $\int_{-L}^{L} f(x) \cos(n\pi x/L)\, dx$ as the area under the curve

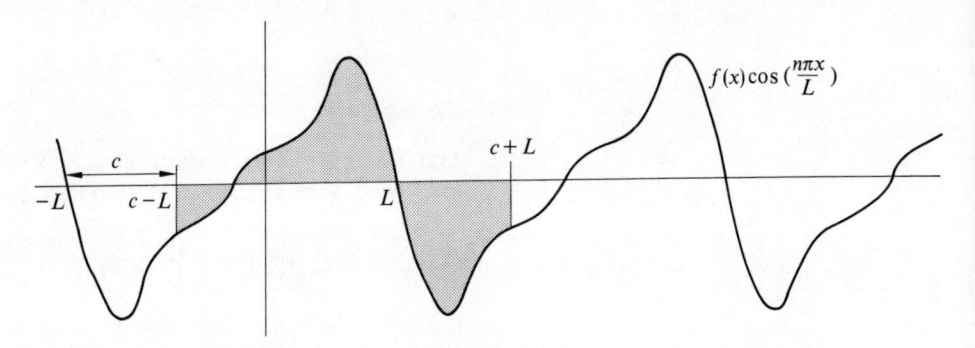

FIGURE 14.5

between $x = -L$ and $x = L$ (figure 14.5) show that a_n may also be written:

$$a_n = \dfrac{1}{L} \int_{c-L}^{c+L} f(x) \cos\left(\dfrac{n\pi x}{L}\right) dx$$

and similarly,
$$b_n = \frac{1}{L}\int_{c+L}^{c+L} f(x) \sin\left(\frac{n\pi x}{L}\right) dx$$
so that it does not matter what range of x we integrate over as long as it is a complete period, $2L$.

8. Sketch the function $f(x) = x^2, 0 < x < 2\pi$, period 2π. Is it odd, even or neither?

Problems 14A

1. Graph each of the following functions over three full periods. Find the Fourier series in each case.

   (i) $f(t) = \begin{cases} 1 & : 0 < t < \pi, \\ -1 & : -\pi < t < 0, \end{cases}$ period 2π

   (ii) $f(x) = x$ $: 0 < x < 2\pi$, period 2π

   (iii) $f(x) = 1 - x^2 : 0 < x < 2\pi$, period 2π

   (iv) $f(t) = \begin{cases} 2 & : 0 < t < 2, \\ 0 & : -2 < t < 0, \end{cases}$ period 4

   (v) $f(x) = x - 3$ $: 0 < x < 6$, period 6

   (vi) $f(x) = x$ $: -4 < x < 4$, period 8

2. Find the points where the functions in problem 1 are discontinuous and find the sum of the series at these points.

3. By putting $x = 2$ in problem 1(vi), show that
$$1 - \tfrac{1}{3} + \tfrac{1}{5} - \cdots = \tfrac{1}{4}\pi$$

4. Deduce from problem 1(iii) that
$$1 - (\tfrac{1}{2})^2 + (\tfrac{1}{3})^2 - (\tfrac{1}{4})^2 + \cdots = \tfrac{1}{12}\pi^2$$

Problems 14B

1. The function $f(x)$ has period 2π and is defined as $f(x) = x \sin x$ for $-\pi < x < \pi$. Show that $f(x)$ may be expressed as:
$$f(x) = 1 - \tfrac{1}{2}\cos x - 2\left\{\frac{\cos 2x}{1.3} - \frac{\cos 3x}{2.4} + \frac{\cos 4x}{3.5} - \cdots\right\}$$

Find by differentiation the Fourier series for the function $f(x) = x \cos x$, $-\pi < x < \pi$, period 2π.

Show that:
$$\tfrac{1}{4} = 1\cdot\tfrac{1}{3} - \tfrac{1}{2}\cdot\tfrac{1}{4} + \tfrac{1}{3}\cdot\tfrac{1}{5} - \cdots$$

and
$$\frac{1}{4} = \frac{3}{2.4} - \frac{5}{4.6} + \frac{7}{6.8} - \cdots$$

2. A function of x of period 2π is equal to $-x^2$ for $-\pi < x \leqslant 0$ and is equal to x^2 for $0 \leqslant x < \pi$. Express the function as a Fourier series, and sketch the graph of the series for values of x between -3π and 5π.

3. Find the Fourier series to represent the function with period 2π defined by

$$f(x) = \begin{cases} x: & 0 \leqslant x < \pi \\ 0: & \pi < x \leqslant 0 \end{cases}$$

4. An alternating current after passing through a rectifier has the form:

$$i = \begin{cases} i_0 \sin \theta: & 0 < \theta < \pi \\ 0 & : -\pi < \theta < 0 \end{cases}$$

and period 2π. Express i as a Fourier series.

5. *Parseval's Identity.* Show that the Fourier series for a function $f(t)$ of period $T = 2\pi/\omega$ is

$$f(t) = \tfrac{1}{2}a_0 + \sum_1^\infty a_n \cos n\omega t + \sum_1^\infty b_n \sin n\omega t \qquad \text{(i)}$$

where

$$a_n = \frac{2}{T} \int_0^T f(t) \cos n\omega t \, dt \quad \text{and} \quad b_n = \frac{2}{T} \int_0^T f(t) \sin \omega t \, dt$$

Multiply (i) by $f(t)$ and integrate to show that:

$$\frac{2}{T} \int_0^T \{f(t)\}^2 \, dt = \tfrac{1}{2}a_0^2 + \sum_1^\infty a_n^2 + \sum_1^\infty b_n^2$$

6. A current is given by $i = f(t)$, where $f(t)$ has period T. Show from Parseval's identity that the r.m.s. value of i is:

$$\sqrt{\left(a_0^2 + 2 \sum_{n=1}^\infty a_n^2 + 2 \sum_{n=1}^\infty b_n^2\right)}$$

where a_n and b_n are the Fourier coefficients. Find the r.m.s. values of the currents
   (i) $i = \text{constant} = \tfrac{1}{2}a_0$ (ii) $i = a_n \cos n\omega t$
   (iii) $i = b_n \sin n\omega t$

7. Show from the results of problem 6 that the power expended by the current i in a resistance R is the sum of the powers expended by the Fourier components taken separately.

15
Half-range Series

In many problems, we are only interested in a function $f(x)$ for values of x in the interval $0 < x < L$, and we are at liberty to define the function as we wish for values of x outside this interval. For instance, we might want to use a Fourier series to represent the shape of a vibrating violin string which is fixed at $x = 0$ and $x = L$. In this case we may define $f(x)$ *outside* $0 < x < L$ as we wish, for we are not interested in the function beyond the points of attachment of the string. But there are two particular ways in which we may define the function outside the interval $0 < x < L$ which give particularly simple Fourier series for $f(x)$. These are:

(i) Define $f(x)$ in $-L < x < 0$, so that $f(x)$ is odd; and then define $f(x)$ for all other x so that it is periodic with period $2L$. In this case, since $f(x)$ is periodic the Fourier series exists, and since it is an odd function all the a_n are zero so that we have a series of sines only. Hence, we are able to represent any function $f(x)$ over $0 < x < L$ as a sine series: this is called a *half-range sine series*.

(ii) Define $f(x)$ to be periodic with period $2L$, and to be an even function. In this case $b_n = 0$ and we have $f(x)$ expressed over $0 < x < L$ as a *half-range cosine series*.

Illustrative Example 1

Expand $f(x) = x$, $0 < x < \pi$ as: (i) a half-range sine series, (ii) a half-range cosine series.

(i) Defining $f(x)$ as odd with period 2π (figure 15.2),

$$a_n = 0, \qquad b_n = \frac{1}{\pi}\int_{-\pi}^{\pi} x \sin nx \, dx$$

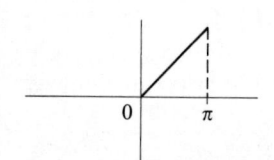

FIGURE 15.1 $y = x, 0 < x < \pi$

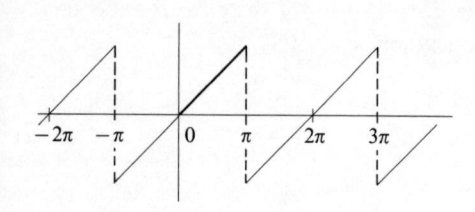

FIGURE 15.2 *Defined as odd, period 2π*

Since the integrand is *even* here, we may write:

$$b_n = \frac{2}{\pi}\int_0^{\pi} x \sin nx \, dx$$
$$= \frac{2}{\pi}\left[\frac{(-x \cos nx)}{n} + \frac{\sin nx}{n^2}\right]_0^{\pi}$$
$$= 2\frac{(-1)^{n+1}}{n}$$

Therefore
$$f(x) = 2(\sin x - \tfrac{1}{2}\sin 2x + \tfrac{1}{3}\sin 3x + \cdots)$$

(ii) Defining $f(x)$ as *even* with period 2π (figure 15.3),

$$b_n = 0, \qquad a_n = \frac{1}{\pi}\int_{-\pi}^{0} -x\cos nx\,dx + \frac{1}{\pi}\int_{0}^{\pi} x\cos nx\,dx$$

FIGURE 15.3 *Defined as even, period 2π*

Therefore
$$a_n = \frac{2}{\pi}\int_{0}^{\pi} x\cos nx\,dx$$
$$= \frac{2}{\pi}\left[\frac{(x\sin nx)}{n} + \frac{(\cos nx)}{n^2}\right]_{0}^{\pi}$$
$$= \frac{2}{\pi}\frac{\{(-1)^n - 1\}}{n^2}$$

Treating a_0 separately,
$$a_0 = \frac{2}{\pi}\int_{0}^{\pi} x\,dx = \pi$$

$$f(x) = \tfrac{1}{2}\pi - \frac{4}{\pi}\left(\cos x + \frac{\cos 3x}{3^2} + \frac{\cos 5x}{5^2} + \cdots\right)$$

We have expressed $f(x)$ as a sine series in (i) and as a cosine series in (ii), but it must be remembered that these two series are only equal for $0 < x < \pi$. Outside this range the two series are entirely different, as we see from their graphs.

Notice that in the above example, both the integrals in the formulae for a_n and for b_n have *even* integrands, so that we were able to write $\int_{-\pi}^{\pi}\ldots = 2\int_{0}^{\pi}\ldots$. In fact this is true generally, since in a half-range sine series the integrand is odd × odd and in a half-range cosine series it is even × even. Hence we may use the following:

Half-range Formulae

Sine series:
$$a_n = 0, \qquad b_n = \frac{2}{L}\int_{0}^{L} f(x)\sin\left(\frac{n\pi x}{L}\right)dx$$

Cosine series:
$$b_n = 0, \qquad a_n = \frac{2}{L}\int_{0}^{L} f(x)\cos\left(\frac{n\pi x}{L}\right)dx$$

Class Discussion Exercises 15

1. Expand $f(x) = \sin x$, $0 < x < \pi$ as a half-range cosine series. Sketch the graph of this series for $-\pi < x < 3\pi$.
2. By differentiating the series obtained in exercise 1, express $f(x) = \cos x$ as a half-range sine series over $0 < x < \pi$. Sketch the graph of the series for $-\pi < x < 3\pi$.
3. Expand the function $f(x) = \begin{cases} 1: & 0 < x < 1 \\ 0: & 1 < x < 2 \end{cases}$
   (i) in a half-range cosine series,
   (ii) in a half-range sine series.
   Sketch the graph of the series for $-3 < x < 5$ in each case.
4B. Show by writing $e^{jx} = \cos x + j \sin x$, $e^{-jx} = \cos x - j \sin x$, that

$$f(x) = \sum_{n=-\infty}^{\infty} c_n \exp\left(\frac{jn\pi x}{L}\right)$$

for a function $f(x)$ of period $2L$, where the coefficients c_n are given by:

$$c_n = \frac{1}{2L} \int_{-L}^{L} f(x) \exp\left(\frac{-jn\pi x}{L}\right) dx$$

Problems 15A

1. Expand $f(x) = \cos x$, $0 < x < \pi$ in a half-range sine series, and graph the sum of the series for $-\pi < x < 3\pi$.
2. Expand

$$f(x) = \begin{cases} 0: 0 < x < \tfrac{1}{2}\pi \\ \pi: \tfrac{1}{2}\pi < x < \pi \end{cases} \quad \text{in a series of sines}$$

3. Expand $f(x) = x(\pi - x)$, $0 < x < \pi$ as a series of (i) cosines, and (ii) sines. Graph each of these series for $-\pi < x < 3\pi$.
   Show that

$$\frac{\pi^2}{6} = 1^2 + (\tfrac{1}{2})^2 + (\tfrac{1}{3})^2 + (\tfrac{1}{4})^2 + \cdots$$

and

$$\frac{\pi^3}{32} = 1^3 - (\tfrac{1}{3})^3 + (\tfrac{1}{5})^3 - \cdots$$

4. Find a Fourier series to represent the function e^x over the half range $0 < x < \pi$ as a sine series and sketch the function represented by the series for $-2\pi < x < 2\pi$. By taking suitable value for x, show that

$$\frac{1}{1+1^2} - \frac{3}{1+3^2} + \frac{5}{1+5^2} - \cdots = \frac{\pi e^{\frac{1}{2}\pi}}{2(e^\pi + 1)}$$

5. If $f(x)$ is an even periodic function with period 2π, defined by $f(x) = x^2$ for $0 \leq x \leq \pi$, sketch its graph for $-3\pi \leq x \leq 3\pi$. Show that the Fourier series for $f(x)$ is

$$f(x) = \tfrac{1}{3}\pi^2 + 4 \sum_{n=1}^{\infty} \frac{(-1)^n \cos nx}{n^2}$$

Obtain a Fourier sine series, valid in the range $0 < x < \pi$ for the function $f(x)$.

6. If

$$f(x) = \begin{cases} x: & 0 < x < \dfrac{\pi}{2} \\ \pi - x: & \dfrac{\pi}{2} < x < \pi \end{cases}$$

show that

$$f(x) = \frac{4}{\pi} \left(\frac{\sin x}{1^2} - \frac{\sin 3x}{3^2} + \frac{\sin 5x}{5^2} - \cdots \right)$$

when $0 < x < \pi$.

7. Expand the function $f(x) = \begin{cases} 1: & 0 < x < \tfrac{1}{2} \\ -1: & \tfrac{1}{2} < x < 1 \end{cases}$

(i) in a half range sine series,
(ii) in a half range cosine series.

Problems 15B

1. If $f(x) = \sin^2 x$ for $0 < x < \pi$, show that its sine series expansion is

$$f(x) = -\frac{8}{\pi} \sum_{n=0}^{\infty} \frac{\sin (2n+1)x}{(2n-1)(2n+1)(2n+3)}$$

2. Find a half-range sine series for the function

$$f(x) = \begin{cases} x: & 0 \leq x < 1 \\ 0: & 1 < x < 2 \end{cases}$$

3. Use Parseval's Identity and illustrative example 1 to show that

$$\frac{\pi^4}{96} = 1 + \frac{1}{3^4} + \frac{1}{5^4} + \cdots$$

Answers to Problems

Problems 1A

1. (i) $\frac{\partial V}{\partial x} = y^2 - 2x$, $\frac{\partial^2 V}{\partial x^2} = -2$, $\frac{\partial V}{\partial y} = 2xy + 2y$,
$\frac{\partial^2 V}{\partial y^2} = 2x + 2$, $\frac{\partial^2 V}{\partial x \partial y} = \frac{\partial^2 V}{\partial y \partial x} = 2y$

(ii) $V_x = 4(x-y)^3$, $V_y = -4(x-y)^3$,
$V_{xx} = V_{yy} = 12(x-y)^2$, $V_{xy} = -12(x-y)^2$

(iii) $z_x = y \cos xy$, $z_y = x \cos xy$,
$z_{xx} = -y^2 \sin xy$, $z_{yy} = -x^2 \sin xy$,
$z_{xy} = \cos xy - xy \sin xy$

(iv) $z_x = e^{x-y}(1+x)$, $z_y = -xe^{x-y}$,
$z_{xx} = e^{x-y}(2+x)$, $z_{yy} = xe^{x-y}$,
$z_{xy} = -e^{x-y}(1+x)$

(v) $\frac{\partial S}{\partial x} = 2x \cos\left(\frac{x}{y}\right) - \frac{x^2}{y} \sin\left(\frac{x}{y}\right)$,
$\frac{\partial S}{\partial y} = \frac{x^3}{y^2} \sin\left(\frac{x}{y}\right)$

(vi) $z_u = v - \frac{1}{u}$, $z_v = u - \frac{1}{v}$, $z_{uu} = \frac{1}{u^2}$, $z_{uv} = 1$,
$z_{vv} = \frac{1}{v^2}$

(vii) $\frac{\partial z}{\partial x} = \frac{y}{x^2 + y^2}$, $\frac{\partial z}{\partial y} = \frac{-x}{x^2 + y^2}$

(viii) $f_x(1,0) = 2$, $f_{xx}(1,1) = 2$, $f_{xy}(x,y) = 0$

(ix) $z_{xx} = \frac{-2y}{x^3}$, $z_{yy} = \frac{2x}{y^3}$, $z_{xy} = \frac{1}{x^2} - \frac{1}{y^2}$

(x) $\frac{\partial z}{\partial x} = \frac{-y}{x\sqrt{(x^2-y^2)}}$, $\frac{\partial z}{\partial y} = \frac{1}{\sqrt{(x^2-y^2)}}$

8. (i) -2 (ii) -2 (iii) 6 (iv) 0

Problems 1B

1. $z_x = \frac{3}{2}$, $z_y = \frac{15}{4}$
6. $n = -\frac{3}{2}$
7. $\left(\frac{\partial x}{\partial r}\right)_\theta = \cos\theta$, $\left(\frac{\partial y}{\partial r}\right)_\theta = \sin\theta$, $\left(\frac{\partial x}{\partial \theta}\right)_r = -r\sin\theta$,
$\left(\frac{\partial y}{\partial \theta}\right)_r = r\cos\theta$, $\left(\frac{\partial r}{\partial x}\right)_y = \frac{x}{r} = \cos\theta$,
$\left(\frac{\partial r}{\partial y}\right)_x = \frac{y}{r} = \sin\theta$, $\left(\frac{\partial \theta}{\partial x}\right)_y = \frac{-y}{r^2} = \frac{-\sin\theta}{r}$,
$\left(\frac{\partial \theta}{\partial y}\right)_x = \frac{x}{r^2} = \frac{\cos\theta}{r}$

Problems 2A

1. 0·31
2. 1 per cent
4. 1/20 m
5. 4·54 m²
6. £32N

Problems 2B

1. (i) $dz = (2xy + y^2) dx + (x^2 + 2xy) dy$

(ii) $dz = \left\{\log\left(\frac{x}{y}\right) + \frac{(x+y)}{x}\right\} dx$
$+ \left\{\log\left(\frac{x}{y}\right) - \frac{(x+y)}{y}\right\} dy$

(iii) $dz = (1 - \sin x) dy - y \cos x \, dx$

(iv) $dz = e^{x+y}(dx + dy)$

(v) $dz = \frac{y}{\sqrt{(1+x^2)}} dx + \sinh^{-1} x \, dy$

2. (i) $2x^3 y$ (ii) $x^2 + 2xy - y^2$
(iii) $x^2 - y^2$ (iv) $y \sin x$

3. (ii) $\frac{1}{3}x^3 + \frac{1}{2}xy^2$ (iii) $-(x+y)^{-1}$
(v) $\sin(x+y)$ (vi) $y - y\cos 2x$

4. $\delta V = -\left(1 + \frac{V}{E}\right)\delta E - \frac{R}{E} \delta R$

Problems 3A

1. (i) $5\sqrt{(5)}/2$ (ii) c (iii) -4 (iv) $13\frac{3}{4}/6$
(v) $\sqrt{2}$ (vi) $-a/\sqrt{2}$

2. (i) $-\sec x$ (ii) $(2-x^2)^{3/2}/x$
(iii) $-\text{cosec } x$ (iv) $\frac{1}{4}x\left(x + \frac{1}{x}\right)^2$

3. (i) $\frac{1}{2}t\left(t^2 + \frac{1}{t^2}\right)^{\frac{3}{2}}$

   (ii) $\frac{1}{2}(4\sin^2\theta + \cos^2\theta)^{\frac{3}{2}}$

   (iii) $3a\sin t\cos t$ (iv) a (v) $-\frac{8}{3}a\cos\frac{t}{2}$

   (vi) $a\theta$ (vii) $-4\cos\frac{t}{2}$

Problems 4A

1. Convergent: (i), (ii), (v). Divergent: (iii), (iv), (vi).
2. Convergent: (i), (ii), (iv). Divergent: (iii).
3. Convergent: (i), (iii), (iv). Divergent: (ii).
4. Series (i) is absolutely convergent.
5. (i) $-1 < x \leqslant 1$ (ii) All x (iii) All x.

Problems 4B

2. Convergent: (i), (ii). Divergent: (iii).
3. (iii) Divergent.

Problems 5A

1. (i) 1 (ii) $\frac{1}{2}$
2. (i) $\dfrac{(n+1)!\,x^n}{1.3\ldots(2n+1)}$, 2 (ii) $\dfrac{(-1)^n x^n}{2^n.(n+1)}$, 2
3. (i) $0.98, 4\times 10^{-4}$ (ii) $0.9804, 8\times 10^{-6}$
   (iii) $1.0204, 8\times 10^{-6}$ (iv) $1.0101, 10^{-6}$
4. -0.051292, remainder. Estimate: 1.6×10^{-6}
5. (i) 1 (ii) 3 (iii) $3^{\frac{3}{4}} \approx 1.316$
6. (i) $\dfrac{x^{2n}}{2^n.n}, \sqrt{2}$ (ii) $\dfrac{n!\,x^{2n-1}}{1.3\ldots(2n-1)}, \sqrt{2}$
7. (i) $0.833333, 8\times 10^{-3}$
   (ii) $0.841667, 2\times 10^{-4}$
   (iii) $0.841468, 3\times 10^{-6}$
8. Up to the term in x^5.
9. 1.25×10^{-7}
10. 1.10517
11. 3

Problems 5B

1. $-2 < x < 0$
2. (i) $-3 < x < 1$ (ii) $-2 < x < 0$
   (iii) $1 < x < 3$ (iv) $x = 1$
   (v) $-4 < x < 0$

3. $0.71945\left(\dfrac{\pi}{180}\right)^2\dfrac{1}{2\sqrt{2}} \approx 1.08\times 10^{-4}$

4. $\left(\dfrac{\pi}{2}\right)^7\cdot\dfrac{1}{7!} \approx 4.7\times 10^{-3}$
5. $0 \leqslant x \leqslant 0.1, 10^{-4}, 1.01\times 10^4$
6. 2.4849

Problems 6A

2. $1 - x + x^2 - x^3 + \cdots \quad -1 < x \leqslant 1$
   $1 + x + x^2 + x^3 + \cdots \quad -1 \leqslant x < 1$
   $1 + x^2 + x^4 + x^6 + \cdots \quad -1 < x < 1$
4. $x - \frac{1}{2}x^2 + \frac{1}{6}x^3 - \frac{1}{12}x^4 + \cdots$ All values of x such that $\sin x \neq -1$, i.e. $x \neq (4n+3)\pi/2$
6. $1 + \dfrac{nx}{1} + n\dfrac{(n-1)}{2!}x^2 + \dfrac{n(n-1)(n-2)}{3!}x^3 + \cdots$

8. (i) 1 (ii) $\frac{1}{3}$ (iii) $\frac{1}{2}$ (iv) $-\frac{1}{2}$ (v) $\dfrac{At}{6l}(1-t^2)$
   (vi) $\frac{2}{3}$ (vii) $\dfrac{t}{4}\sin 2t$ (viii) $\frac{4}{3}$ (ix) 1

9. 0.0201
10. $x - \frac{1}{3}x^3 + \frac{2}{15}x^5 + \cdots$

Problems 6B

1. $\frac{7}{360}$
3. $\dfrac{E_m t \sin nt}{2L}$
5. 0.69315
6. 0.0719
7. $x - \frac{1}{3}x^2 + \frac{2}{15}x^5$
8. 2.2361
9. 5000 (approx.); 9
10. $\lim_{x\to 0}\dfrac{y^2}{2x}$ (i) -1 (ii) 4
11. 0.0196

Problems 7A

1. (i) $-e^{-x}(x^2 - 6x + 6)$
   (ii) $8x\cos 2x + 16\sin 2x$
   (iii) $-2e^x(\cos x + \sin x)$
   (iv) $(x^2 - 20)\cos x + 10x\sin x$
   (v) $e^x(x^2 + 2nx + n(n-1))$
2. $-2xe^{-x^2}, (4x^2 - 2)e^{-x^2}$

Problems 7B

2. $a\left(x + \dfrac{1^2 - a^2}{3!}x^2 + \dfrac{(1^2 - a^2)(3^2 - a^2)}{5!}x^5 + \cdots\right)$

3. $2\left(\dfrac{x^2}{2} + \dfrac{2}{3}\cdot\dfrac{x^4}{4} + \dfrac{2\cdot 4}{3\cdot 5}\cdot\dfrac{x^6}{6} + \dfrac{2\cdot 4\cdot 6}{3\cdot 5\cdot 7}\cdot\dfrac{x^8}{8} + \cdots\right)$

Problems 8A

1. (i) $\cos t - \tfrac{1}{3}\cos^3 t$ (ii) $\tfrac{2}{15}$
   (iii) $\log|1 + \sin 3x|$ (iv) $\tfrac{1}{8}x - \tfrac{1}{32}\sin 4x$
   (v) $x - \tanh x$ (vi) $\log|\sinh x|$
2. (i) $\tfrac{1}{2}\sec^2\theta + \log|\cos\theta|$ (ii) $2\sqrt{2}\sin(x/2)$
   (iii) $\tfrac{1}{4}\cos 2x - \tfrac{1}{24}\cos 12x$ (iv) $\tfrac{2}{3}(\sin x)^{\frac{3}{2}}$
   (v) $\sinh\theta + \tfrac{2}{3}\sinh^3\theta + \tfrac{1}{5}\sinh^5\theta$
   (vi) $\sqrt{2}\sinh x$
3. (i) $\tfrac{1}{4}\sinh 2u + \tfrac{1}{2}u$ (ii) $\tfrac{1}{4}\sinh 2u - \tfrac{1}{2}u$
   (iii) $\tfrac{1}{6}\sinh^6 t$ (iv) $2\sqrt{2}\cosh\tfrac{1}{2}x$
4. (i) $\tfrac{1}{4}\log|(2 + \tan\tfrac{1}{2}\theta)/(2 - \tan\tfrac{1}{2}\theta)|$
   (ii) $\log|\tan\tfrac{1}{2}\theta|$ (iii) $-2/(1 + \tan\tfrac{1}{2}\theta)$
   (iv) $\tfrac{1}{2}\log(2\sqrt{3} - 1)$ (v) $2\pi/3\sqrt{3}$
5. (i) $0, i_1/\sqrt{2}$ (ii) $2/\pi, 1$
   (iii) $0, \sqrt{[\tfrac{1}{2}(i_1^2 + i_2^2)]}$ (iv) $i_0, \sqrt{(i_0^2 + \tfrac{1}{2}i_1^2)}$
6. (i) $\dfrac{1}{a}\log(2 + \sqrt{3})$ (ii) $(1 + \tfrac{1}{2}\pi)/32$
   (iii) $\log(1 + \sqrt{2}) - 1/\sqrt{2}$ (iv) $\tfrac{1}{3}(2\sqrt{2} - 1)$
   (v) $\tfrac{1}{4}\pi a^2$
   (vi) $\tfrac{1}{2}[\sqrt{2} + \sinh^{-1} 1]$ or $\tfrac{1}{2}[\sqrt{2} + \log(1 + \sqrt{2})]$
7. (i) $\tfrac{1}{2}[x(x^2 - 9)^{\frac{1}{2}} - 9\cosh^{-1}\tfrac{1}{3}x]$
   (ii) $\tfrac{3}{4}\sqrt{13}$ (iii) $-\tfrac{1}{4}(4 - x^2)^{\frac{1}{2}}/x$
   (iv) $4 - 3\cos^{-1}(\tfrac{3}{5})$ (v) $x/(1 - 4x^2)^{\frac{1}{2}}$
   (vi) $\tfrac{1}{6}[3x\sqrt{(9x^2 + 1)} + \sinh^{-1} 3x]$
11. $k = 4/\pi a^2$, $\mu = 4a/3\pi$, $\sigma^2 = \tfrac{1}{4}a^2$

Problems 8B

1. (i) $2\log|(1 + \tan\tfrac{1}{2}x)/(1 - \tan\tfrac{1}{2}x)| - \tan\tfrac{1}{2}x$
   (ii) $\log|\tan\tfrac{1}{2}x| + 2/(1 + \tan\tfrac{1}{2}x)$
2. (i) $\dfrac{1}{\sqrt{2}}\log|\tan\tfrac{1}{2}\theta|$ (ii) $-\log|1 + \cos^2 x|$
   (iii) $\tfrac{1}{4}\tan^4 x - \tfrac{1}{2}\tan^2 x + \log|\sec x|$
   (iv) $\log|\cosh\theta| + \tfrac{1}{2}\text{sech}^2\theta$
   (v) $\tfrac{1}{2}\tanh^2\theta$ or $-\tfrac{1}{2}\text{sech}^2\theta$
   (vi) $\sqrt{2}\log|\tanh\tfrac{1}{4}\theta|$
3. $6a$
4. $\log(1 + \sqrt{2})$
5. $4ab\pi/3\alpha + (a^2 + b^2)\pi^2/4\alpha$

7. (i) $\tfrac{1}{2}\{3\sqrt{10} - 2\sqrt{5} + \log\{(\sqrt{10} + 3)/(\sqrt{5} + 2)\}\}$
   (ii) $\dfrac{\sqrt{3}}{2} - \dfrac{1}{4}\log(2 + \sqrt{3})$
8. (i) $\tfrac{1}{2}[\tan^{-1} u + u/(1 + u^2)]$ (ii) 1
   (iii) $-\tfrac{1}{2}\log(\tan\tfrac{1}{8}\pi)$
9. $2\pi - 2$. (Hint: multiply top and bottom of integrand.)
10. $\pi/6$ (Hint: substitute $u = x^2$)
11. $\tfrac{3}{8}[2\sqrt{3} - \tfrac{2}{3} + \log(1 + 2/\sqrt{3})]$

Problems 9A

1. (i) $3\pi/16$ (ii) $16/35$
2. $24(1 - e^{-\pi})/85$
3. (i) $-\tfrac{1}{5}\cot^5 x + \tfrac{1}{3}\cot^3 x - \cot x - x$
   (ii) $-\tfrac{1}{6}\cot^6 x + \tfrac{1}{4}\cot^4 x - \tfrac{1}{2}\cot^2 x - \log|\sin x|$
4. $I_n = -x^n e^{-x} + nI_{n-1}$. Integrand is positive and less than 1 over range of integration. Hence consider integral as an area ($I_5 \approx 0\cdot 1$).
5. $I_n = \dfrac{1}{n^2} + \dfrac{n-1}{n}\cdot I_{n-2}$, $n \geq 2$; $\tfrac{1}{4} + 3\pi^2/64$
7. $32a^{\frac{3}{2}}/315$

Problems 9B

1. (i) $\tfrac{5}{12} - \tfrac{1}{2}\log 2$ (ii) $\tfrac{1}{5}x^5 - \tfrac{1}{3}x^3 + x - \tan^{-1} x$
2. $I_n = x^n(2x + c)/(2n + 1) - nc\, I_{n-1}/(2n + 1)$; $2(7 - \sqrt{3})/35$
3. (i) $\tfrac{2}{15}$ (ii) $\pi/32$ (iii) $\tfrac{1}{18}$
5. $81/140$
6. $I_n = x(\log_e x)^n - nI_{n-1}$

Problems 10A

1. (i) $x^2 + y^2 = a^2$ (ii) $x^2 + y^2 - x - y = 0$
   (iii) $x + y = 0$ (iv) $y = 1$
   (v) $x^2 + (y - a)^2 = a^2$
   (vi) $x = 0$
2. (i) $r = 2a\sin\theta$ (ii) $r(\cos\theta - \sin\theta) = 1$
   (iii) $r^2 = a\tan\theta$ (iv) $\theta = \alpha$
3. (i) Equiangular spiral (see problem 9).
   (ii) Symmetric about Ox. Cardioid with cusp at $r = 0$, $\theta = 0$.
   (iii) Symmetric about Ox. $r = \infty$ when $\theta = 0$. Parabola.
   (iv) Parabola.

(v) Symmetric about Ox.
(vi) Straight line.
(vii) Symmetric about Ox.
(viii) Symmetric about Ox and origin.
(ix) Symmetric about Oy.
(x) Symmetric about Ox.
(xi) Cardioid.
(xii) Cardioid.

4. $3\pi a^2/2$
5. 6π
6. (i) $\frac{1}{2}\int_{-\pi/3}^{\pi/3} 2^2\, d\theta - \sqrt{3} = 4\pi/3 - \sqrt{3}$

    (ii) $\frac{1}{2}\int_{-\pi/3}^{\pi/3} [4(1+\cos\theta)/3]^2\, d\theta - \sqrt{3}$
    $= 8\pi/9 + 20\sqrt{3}/9$

7. (i) $2\int_0^{\pi/4} \frac{1}{2}\cos^2 2\theta\, d\theta = \frac{1}{8}\pi a^2$
    (ii) $\frac{1}{2}a^2$
8. (i) $8a$ (ii) $2\pi a$
    (iii) $2\pi a$ (iv) 4
10. $279°$

Problems 10B

1. $\operatorname{Tan}^{-1}(-3\sqrt{3})$
2. $2\int_0^{2\pi/3} \frac{1}{2}r^2\, d\theta = 2\pi + 3\sqrt{(3)}/2$; $2\int_{2\pi/3}^{\pi} \frac{1}{2}.r^2\, d\theta$
    $= \pi - 3\sqrt{(3)}/2$
3. $(2+\frac{1}{4}\pi)a^2$
4. $\pi/3 + \frac{1}{2}\sqrt{3}$
5. $\frac{3}{8}\pi$
6. $9a^2/128$, $\frac{1}{2}a(\frac{15}{16} + \log 2)$

Problems 11A

1. $(\frac{2}{3}h, \frac{1}{3}mh)$; $\frac{1}{3}h\pi r^2$
2. $2^5\pi^2$
4. (i) $(\frac{5}{7}, \frac{5}{16})$; $\frac{1}{4}\pi$ (ii) $(\frac{9}{5}, \frac{9}{8})$; $168\pi/5$
    (iii) $(0, \frac{3}{5})$; $8\pi/5$
    (iv) $\left(\dfrac{e+1}{4}, \dfrac{1}{e-1}\right)$; $2\pi, \frac{1}{2}\pi(e^2-1)$
    (v) $(\frac{1}{2}, \frac{1}{5})$ (vi) $(\frac{3}{7}, \frac{12}{25})$
5. $(4a/3\pi, 4a/3\pi)$, $2\sqrt{2\pi a^3/3}$
6. (i) $(2, 0)$ (ii) $(0, \frac{5}{4})$ (iii) $\bar{x} = \frac{4}{5}$
    (iv) $\bar{x} = \frac{3}{4}h$ (v) $\bar{y} = \frac{1}{2}(e^2+1)/e^2-1)$
7. $\frac{1}{2}a$
8. $\frac{7}{9}$

Problems 11B

1. $\bar{x} = \bar{y} = 2\cdot 7$, $V_{Ox} = 57\cdot 8\pi$
2. $V_{Ox} = \pi^2/4$, $V_{Oy} = \pi^2 - 2\pi$, $\bar{x} = \frac{1}{2}\pi - 1$, $\bar{y} = \frac{1}{8}\pi$
3. $\frac{3}{16}\pi a$
4. $\bar{x} = \frac{58}{35}$; $232\pi/15$
5. By symmetry, $\bar{x} = \pi a$, $\bar{y} = \frac{2}{3}a$.
6. $\bar{x} = \bar{y} = \frac{2}{5}a$
7. Symmetrically placed, $\dfrac{2}{3}\dfrac{\sin\alpha}{\alpha}$ from vertex

Problems 12A

2. $2Ma^2$
3. $\sqrt{\frac{5}{6}}$
4. (i) $2Ma^2$ (ii) $\frac{1}{2}Ma^2$
5. $\frac{1}{3}Ma^2$
6. (i) $\frac{5}{3}Ma^2$ (ii) $\frac{10}{3}Ma^2$
7. $\frac{1}{2}Mh^2$
8. (i) $6\cdot 48$ (ii) $2\cdot 87$
9. $\frac{2}{3}Ma^2$, $\frac{2}{5}Ma^2$
10. $9\cdot 53$

Problems 12B

3. $\frac{3}{4}\pi\rho$
4. $\sqrt{\frac{2}{5}}$

Problems 13A

1. (i) $0\cdot 693$ (ii) $1\cdot 00$ (iii) $0\cdot 386$ (iv) $0\cdot 347$
    (v) $1\cdot 27$ (vi) $0\cdot 481$
2. $2\cdot 05$; By expansion: $2\cdot 047$
3. $2\cdot 58$ (3 significant figures)
4. $0\cdot 5236$; $\pi \approx 3\cdot 142$
5. $0\cdot 508$
6. $43\cdot 63$
7. $0\cdot 592$

Problems 13B

2. 242 m
3. $0\cdot 592 u/g$
4. $-\frac{1}{3}$, exact
5. 40 strips
6. 2 decimal places, 6 decimal places

Problems 14A

1. (i) $\dfrac{4}{\pi}(\sin t + \tfrac{1}{3}\sin 3t + \tfrac{1}{5}\sin 5t + \cdots)$

   (ii) $\pi - 2\sum_{n=1}^{\infty}\dfrac{\sin nx}{n}$

   (iii) $1 - \tfrac{4}{3}\pi^2 - \sum_{n=1}^{\infty}\left(\dfrac{4}{n^2}\cos nx - \dfrac{4\pi}{n}\sin nx\right)$

   (iv) $1 + \dfrac{4}{\pi}\left(\sin\dfrac{\pi t}{2} + \tfrac{1}{3}\sin\dfrac{3\pi t}{2} + \tfrac{1}{5}\sin\dfrac{5\pi t}{2} + \cdots\right)$

   (v) $-\dfrac{6}{\pi}\sum_{n=1}^{\infty}\dfrac{1}{n}\sin\tfrac{1}{3}n\pi x$

   (vi) $\dfrac{8}{\pi}\left(\sin\tfrac{1}{4}\pi x - \tfrac{1}{2}\sin\dfrac{2\pi x}{4} + \tfrac{1}{3}\sin\dfrac{3\pi x}{4} - \cdots\right)$

Problems 14B

1. $-\tfrac{1}{2}\sin x + 2\left(\dfrac{2}{1\cdot 3}\sin 2x - \dfrac{3}{2\cdot 4}\sin 3x + \dfrac{4}{3\cdot 5}\sin 3x - \cdots\right)$

2. $f(x) = \sum_{n=1}^{\infty}\dfrac{2\pi(-1)^{n+1}}{n} + \dfrac{4}{\pi n^3}[(-1)^n - 1]\sin nx$

3. $\tfrac{1}{4}\pi - \dfrac{2}{\pi}\sum_{n=0}^{\infty}\dfrac{\cos(2n+1)x}{(2n+1)^2} + \sum_{n=1}^{\infty}(-1)^{n+1}\dfrac{\sin nx}{n}$

4. $i = i_0\left[\dfrac{1}{\pi} + \tfrac{1}{2}\sin\theta - \dfrac{2}{\pi}\sum_{n=1}^{\infty}\dfrac{\cos 2n\theta}{4n^2 - 1}\right]$

Problems 15A

1. $\dfrac{8}{\pi}\sum_{n=1}^{\infty}\dfrac{n\sin 2nx}{4n^2 - 1}$

2. $2\sum_{n=1}^{\infty}(-\cos n\pi + \cos\tfrac{1}{2}n\pi)\dfrac{\sin nx}{n}$

3. (i) $\dfrac{\pi^2}{6} - \left[\dfrac{\cos 2x}{1^2} + \dfrac{\cos 4x}{2^2} + \dfrac{\cos 6x}{3^2} + \cdots\right]$

   (ii) $\dfrac{8}{\pi}\left[\dfrac{\sin x}{1^3} + \dfrac{\sin 3x}{3^3} + \dfrac{\sin 5x}{5^3} + \cdots\right]$

4. $\dfrac{2}{\pi}(1 + e^\pi)\left[\dfrac{\sin x}{1 + 1^2} + \dfrac{\sin 3x}{1 + 3^2} + \dfrac{\sin 5x}{1 + 5^2} + \cdots\right]$

5. $\sum_{n=1}^{\infty}\left[2\pi\dfrac{(-1)^{n+1}}{n} + \dfrac{4}{\pi n^3}\{(-1)^n - 1\}\right]\sin nx$

7. (i) $\dfrac{4}{\pi}\sum_{n=0}^{\infty}\dfrac{\sin 2(2n+1)\pi x}{(2n+1)}$

   (ii) $\dfrac{4}{\pi}\sum_{n=1}^{\infty}(-1)^{n+1}\dfrac{\cos(2n-1)\pi x}{(2n-1)}$

Problems 15B

2. $\sum_{n=1}^{\infty}\left\{-\dfrac{2}{n\pi}\cos\dfrac{n\pi}{2} + \dfrac{4}{n^2\pi^2}\sin\dfrac{n\pi}{2}\right\}\sin\dfrac{n\pi x}{2}$

Index

Absolute convergence, 29
Alternating series, 29
Alternator, 41
Arc length in polars, 61
Area in polars, 61
Area of revolution in polars, 64
Average, 19

Binomial series, 37

Cardioid, 63, 64
Centre of curvature, 21
Centre of mass, 65
Centroid, 66
Circle of curvature, 21
Comparison test, 27
Convergence, 26–31
Capacitance, 41
Curvature, 21–25

D'Alembert's ratio test, 28
Density function, 53
Derivative, partial, 11
—, higher order, 12
—, of a product, 43
Differentiation, partial, 11
—, of series, 38
Divergent series, 26

Evolute, 22, 24

Fourier series, 81–87
—, in complex exponential form, 90

Gradient of a curve, 12, 14

Half-range series, 88
Harmonic series, 29
Hyperbolic substitutions in integrals, 48

Integration of series, 38

Laplace's equation, 14
Leibniz' rule for derivative of a product, 43–46
Leibniz' theorem for convergence, 29

Mean value, 52, 53
Moment of inertia, 70
—, table of, 73
Moment of area, 66

Numerical integration 75–80

Osborne's rule, 47, 50

Pappus' theorems, 67
Parallel axes theorem, 71
Parseval's identity, 87, 91
Partial derivative, 11
Period of a compound pendulum, 17
Periodic functions, 81
Perpendicular axes theorem, 71
Polar coordinates, 59–64
Power, 16
Probability density function 53

Radius of convergence, 32–36
—, Newton's formula, 42
Radius of curvature, 21
Radius of gyration, 70
Ratio test, 28
Reduction formulae, 55–58
Remainder after n terms of a series, 33
Resistor, 16
Root mean square, 52

Series expansions of well-known functions, 37
Simpson's rule, 77
—, estimate of error, 78
Small oscillations, 17
Summable series, 26
Spiral, 62

Toroid, 74
Total differential, 17
Trapezoidal rule, 76
Trigonometric substitutions in integrals, 48
Trigonometric identities, 47

Variance, 18, 53

Wallis' formulae, 56–58